Mankind's Problem: Climate Change

Albert Fässler

Mankind's Problem: Climate Change

Scientific Facts with Philosophical Considerations and Contributions

Translated by Dr. Baoswan Dzung Wong

 Springer

Albert Fässler
Berner Fachhochschule
Bern, Switzerland

ISBN 978-3-662-71845-2   ISBN 978-3-662-71846-9   (eBook)
https://doi.org/10.1007/978-3-662-71846-9

Preface

A thick book is a great evil.

Gotthold Ephraim Lessing

Motivation for and content of this book

My lecture entitled *A Mathematical Model for Climate Change* at the Swiss Federal Institute of Technology, ETH Zurich in Switzerland in December 2021 on the invitation of Prof. Norbert Hungerbühler triggered positive reactions throughout. The heterogeneous audience consisted of engineers, physicians, physicists, mathematicians, computer scientists, teachers as well as a psychologist and a pilot. Their feedback motivated me to write the book with title *Menschheitsproblem Klimaänderung, Wissenschaftliche Tatsachen mit Philosophischen Betrachtungen und Beiträgen, Springer Verlag, 2024.*
Even before the original German edition was published, there were numerous requests for an English translation, which is now available.

The lecture was based on parts of the Climate Change Section from my book *Schnelleinstieg Differentialgleichungen, 2. Auflage, Springer Verlag, 2020.* English translation: *Fast Track to Differential Equations, second edition, Springer, 2021.*

The problem of climate change is extremely interdisciplinary and reflects the complexity of our planet Earth.

This book deals with a simple, well-established mathematical model, based on scientific principles, and deals with the causes and effects. In addition, it gives an overview of relevant facts and data, compiled from a full range of sources. Finally, it was important to me to also bring in the views of philosophers on the subject.
It is aimed at all interested parties, in particular economic and political decision-makers. It also addresses readers who have little relation to mathematics and have little or no knowledge of the concept of differential equations. It is introduced gently, in a simple manner in the appendix. It is my intention to enable an in-depth, differentiated and factual formation of opinions on the climate problem.

The computation of future temperature profiles for different scenarios in Section 2.2 can be ignored if one is satisfied with the graphic solutions.

I tried to describe the content according to Albert Einstein's view: *Everything should be explained as simply as possible, but not simpler.*

V

There is No Plan B for the Earth

The unconceivably large universe is about 14 billion $= 1.4 \cdot 10^{10}$ years old and contains an incredibly large number of celestial bodies with billions of galaxies, each of which contains billions of stars.

The following image taken from [69] shows the heart of the spiral galaxy Messier M74[1] (Phantom Galaxy M74) and was taken with NASA's James Webb Telescope.

It is about 32 million light-years[2] away from us and has a radius of about 47,500 light-years.

The Hickson Compact Group 40[3] includes a quintet of galaxies at a distance of even as far as about 300 million light-years. Such astronomical data trigger modesty and humility.

To conclude from the gigantic abundance of celestial bodies that there is a planet B for us is absurd. The average distance to Mars, the nearest planet, but hostile to life, is about 70 million kilometers. We only have planet A, our earth!

In 1941, the German climatologist and meteorologist Hermann Flohn (1912–1997) already put forward the thesis of man-made climate change: "Human activity is the cause of global climate change, the future significance of which nobody can foresee."

As early as 1972, the *Club of Rome* caused a sensation with its book *Limits to Growth* with topics of population explosion, shortage of raw materials and a destroyed environment.

---

[1] Discovered in 1780 by French astronomer Pierre Méchain, assistant to Charles Messier.

[2] 1 light-year is the distance that light travels at a speed of 300,000 km/s in one year, i.e. 9.46 trillion kilometers $\approx 10^{13}$ km.

[3] Paul Hickson, born 1950, is a British-Canadian astronomer. In 1982 he published a catalog of 100 galaxy groups.

From today's perspective, the sober statement applies that there is a blatant contradiction between knowledge and action. 2024 was the warmest year on record!

From [3] *Submission: We are not the masters of the world.* "Man sees himself as a being that stands outside of nature and he regards the world as his property. A story of domination over nature and human arrogance. We can imagine many things. But not a world without people. The thought that the species to which we belong could be a mere episode in the history of the planet is unbearable to us, even though we know full well that other inhabitants of the earth are becoming extinct, ongoing at incredible rates. 150 animal and plant species are disappearing every day. And there is no end in sight. But for humans it always has to go on, somehow. Of course, different rules apply to them."

This matches the excellent movie [64] entitled *Carbon, A Story of Life and Death.* No life without carbon. But too much CO_2 in the atmosphere from burning fossils is fatal. The movie ends with the statement of a personified carbon atom: "Whatever you humans will do: I shall survive."

About this book

It includes updated and enhanced material compared to the German original entitled „Menschheitsproblem Klimaänderung."

I owe Dr. Baoswan Dzung the excellent translation into English, including editorial as well as graphical improvements. I had the pleasure to study at ETH Zürich with her. She continued her studies at the University of California, Berkeley, U.S.A and finished with a Ph.D. in the field of functional analysis under the guidance of Professor Tosio Kato.

With her precise work style and her highly professional abilities, she improved this edition from the mathematical and from the linguistic point of view.

I wish to express my big thanks to her for the beautiful translation and the inspiring interaction during the process of materializing this book.

Acknowledgments

My sincere gratitude goes to Dr. Nicolas Gruber, Professor of Environmental Physics at the Department of Environmental Sciences at ETH Zurich for his extremely valuable support and active participation in analyzing the model.

I would like to thank Dr. Reto Knutti, Professor of Climate Physics at ETH Zurich, long-standing member of the Intergovernmental Panel on Climate Change IPCC, for his input and critical answers to my questions.

Prof. Pierre Friedlingstein, Director of the Global Carbon Budget Office at the University of Exeter, gave me scientific instructions.

The glaciologists Prof. Daniel Farinotti and Dr. Mauro Werder, both at ETH Zurich, gave me support on the subject of glaciers.

Thanks go to Dipl. Phys. Veronika Erdmann, Associate Editor for Mathematics and Statistics and to Anja Groth, Project Manager from Springer Publishing for the pleasant cooperation.

A thank-you goes to Dr. Pankaj Shrivastasa[4], who organized a video conference in India in 2023 for my talk titled *Climate Change*.

A *merci* goes to Olivier Fässler for solving my computer problems and to Markus Fässler for his perfect environnment at the Lake of Constance in the peak of summer heat in 2025, where I could work on the lengthy completion of the book.

I owe improvements of certain graphics to Dr. Walter Businger.

I am grateful to my wife Carmen Fässler, who is interested in the subject, for providing me with a pleasant, quiet working environment and some of the press references.

Dear reader:
The mankind problem of climate change is decidedly interdisciplinary. It combines climatology, physics, chemistry, biology, environmental sciences, geography, glaciology, ecology, economy and sociology.

I have tried to describe essential aspects of the global problem in a way that is easy to understand and I have tried to identify some current data from the abundance of information from relevant sources with the aim of presenting to you an overview of the diverse topic.

The computation and compilation of my own diagrams were performed by the computer algebra system *Mathematica*.

CH-2533 Evilard, August 2025,
Albert Fässler.

---

[4] Professor at the M.N. National Institute of Technology in Uttar Pradesh, Secretary General of the Forum for Advanced Training in Education and Research, Academy of India (FAI).

Contents

Chapter 1
Solar and terrestrial thermal radiation

Electromagnetic radiation is the transmission of energy by means of electromagnetic waves or photons.

Every body emits and absorbs electromagnetic radiation. A body that absorbs and emits the entire incident radiation over all wavelengths (i.e. over the entire spectrum) is called an **ideal black body** or simply a **black body.** It is an idealized source of thermal radiation. Intensity and spectral distribution of its thermal radiation depend only on its temperature and are therefore independent of the type of body and its surface.

For a black body, **Planck's law of radiation** describes the intensity distribution of the electromagnetic power as a function of the wavelength λ.

The radiation spectrum as a function of the wavelength λ consists of three ranges, with the boundaries not being precisely defined. To describe them we use the usual units:

1 nanometer = 1 nm = 10^{-9} m, 1 micrometer = 1μm = 10^{-6} m = 1,000 nm.

1. Short-wave **ultraviolet radiation (UV radiation)** includes wavelengths from 100 nm to 400 nm.
2. The **visible light** covers the range from 400 nm (violet) across blue, green, yellow to 780 nm (red).
3. Long-wave **infrared radiation (IR radiation)** covers the range from 780 nm to 1 millimeter.

To a good approximation, the sun and the earth can be viewed as black bodies.

- The sun primarily emits short-wave radiation.
- The earth radiates the absorbed solar radiation into the atmosphere as a long-wave infrared radiation.

A. Fässler, *Mankind's Problem: Climate Change,*
https://doi.org/10.1007/978-3-662-71846-9_1

In the diagram below, the two spectra are plotted as functions of the wavelength λ measured in micrometers.

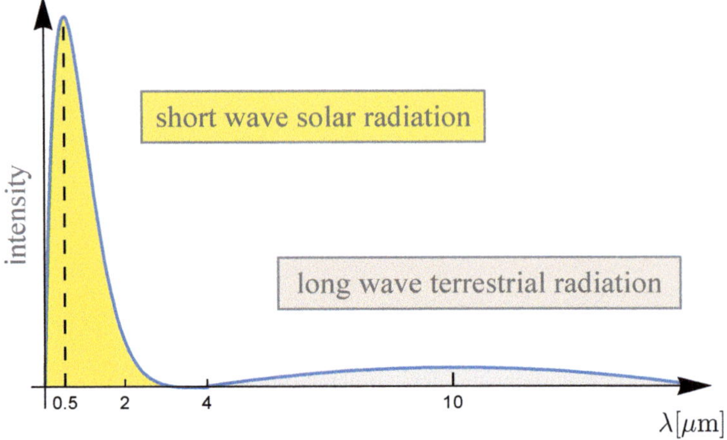

The maximum radiation intensity of the sun occurs at a wavelength of about 0.5 μm, that of the earth at about 10 μm.

Wien's displacement law[1] states that the wavelength at which a black body with the absolute temperature T emits the most intense radiation is inversely proportional to T.

If the sun has an approximate temperature of 5,900 K and the earth that of 287 K, then Wien's displacement law implies

$$\frac{10}{0.5} \approx \frac{5900}{287}.$$

Wavelength λ and frequency f are inversely proportional to each other:

$$c = \lambda \cdot f.$$

Here $c \approx 300,000$ km/s is the speed of light in a vacuum. The frequency has the dimension s^{-1} and is denoted by Hertz Hz.

For the most intensive solar radiation one has

$$\lambda = 0.5 \ \mu m \quad \text{and} \quad f = \frac{3 \cdot 10^8 \, \text{m/s}}{0.5 \cdot 10^{-6} \text{m}} = 6 \cdot 10^{14} \, \text{Hz}.$$

The *total mean solar radiation (Total Solar Irradiation TSI)* onto a surface perpendicular to the sun's rays outside the atmosphere is

$$S_0 = 1370 \frac{\text{W}}{\text{m}^2}.$$

It has the dimension watts per square meter, i.e. the dimension of the power output per square meter, and is equivalent to the area under the solar spectral curve.

---

[1] Wilhelm Wien, German physicist, born 1864 in Russia, died 1928 in Germany.

Absorption, reflection and dispersion of the solar radiation determine the power received on the earth's surface.

The following figure[2] shows three spectra of solar radiation: that of the ideal black body and the spectra of extraterrestrial and terrestrial solar radiation. There are also gaps, caused by resonance excitation of certain frequencies of molecules, which manifest themselves in thermal movements.

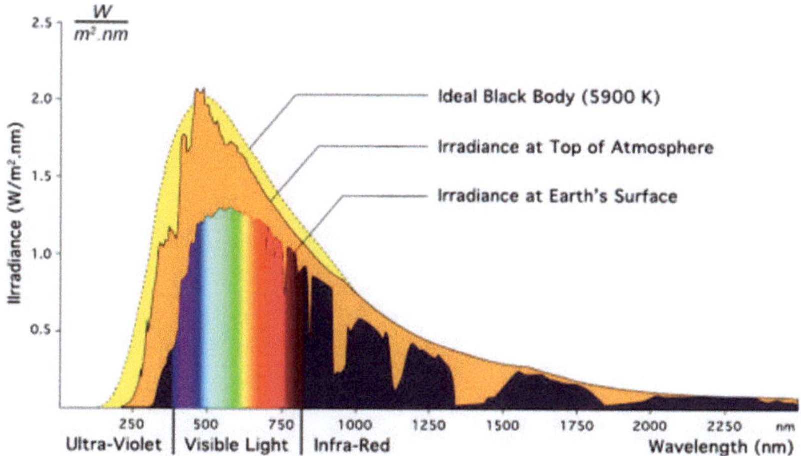

The ozone layer, which is at an altitude of about 15 km to 30 km, absorbs part of the harmful UV radiation:

1. Only the longer-wave UV-A radiation (400–315 nm) passes through the ozone layer practically unfiltered.
2. About 10% of the UV-B radiation (315–280 nm) reach the earth's surface;
3. and the UV-C radiation (280–100 nm) is almost completely blocked.

An impressive comparison, taken from [7]:

- **Mean power consumed by the world population.**
  The average energy consumption per day and citizen is 58 kWh, which is equivalent to an average power of 2.4 kW.
  The worldwide population of 8 billion people consumes an average power of about $8 \cdot 10^9 \cdot 2.4\,\text{kW} \approx 1.9 \cdot 10^{13}\,W = 19$ Terawatt $= 19$ TW.
  $1\text{T} = 1$ Tera $= 10^{12}$.
- **Average power of solar radiation on earth.**
  The terrestrial radiation is estimated at roughly 70% of the extraterrestrial radiation. This is evident when comparing the areas of the two spectral distribution functions in the figure above.
  With the radius of the earth being $R = 6,371$ km it amounts to about $0.7 \cdot S_0 \cdot \pi R^2 \approx 1.2 \cdot 10^{17}\,W = 120,000$ TW.

The power emitted by the sun is therefore 6,000 times the power consumed by the world's population!

---

[2] from Science Direct: www.sciencedirect.com

Chapter 2
The zero-dimensional energy balance model

This model was developed by two individuals:[1]

- The Russian climatologist, geophysicist and geographer Mikail Ivanovich Budyko (1920–2001). He was considered one of the leading climate researchers. Several models and predictions for global warming go back to his work.

- The American-British climate researcher and astronaut Piers John Sellers (1955–2016). After his three space shuttle flights between 2002 and 2010, he became a Director of the Earth Science Research Department at NASA's Goddard Space Flight Center.

---

[1] The two portraits are taken from Wikipedia.

2.1 Differential equation and temperature limits

The physical model is based on the balance of heat output: the difference between the heat gain P_{gain} absorbed by the earth via solar radiation and the heat P_{loss} lost via radiation results in the earth heating up.
All physical quantities in this chapter are **exclusively globally averaged scalar quantities**, which is why we speak of a zero-dimensional model.

The quantity of heat of a body of mass M with absolute temperature T measured in Kelvin and specific heat capacity C is

$$Q = C \cdot M \cdot T. \tag{2.1}$$

In the case of the earth, C is the *specific mean global heat capacity* of the affected layer of the earth's surface per degree and kg.

The time-dependent rate of change in the quantity of heat per unit of time $\frac{dQ}{dt}$ (physically a power) is described by the following **heat equation** of a heat-exchanging body of arbitrary shape:

$$\frac{dQ}{dt} = P_{\text{gain}} - P_{\text{loss}}. \tag{2.2}$$

The total constant mean power of the short-wave solar radiation absorbed by the earth is given by

$$P_{\text{gain}} = \pi R^2 \cdot S_0 \cdot (1 - \alpha).$$

Here πR^2 is the cross-sectional area of the earth with radius R. Moreover, $\alpha \approx 0.32$ is the portion of the reflected radiation, the so-called **planetary albedo**, that is, 32% of the solar radiation is reflected.

Why a circular disk? The sun's radiation on the small surface elements of half the sphere's surface is in general oblique, not perpendicular. Therefore, only the surface area of the illuminated sphere projected onto the plane perpendicular to the irradiation counts for the power.

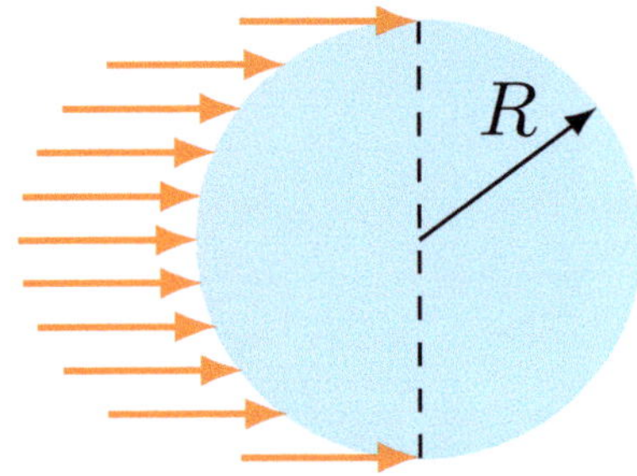

The total power loss caused by the outgoing long-wave radiation is given by the **Stefan-Boltzmann law**[2]

$$P_{\text{loss}} = 4\pi R^2 \cdot \varepsilon \cdot \sigma \cdot g \cdot T^4.$$

Here

- $\varepsilon = 0.97$ is the emissivity, which specifies how much radiation the earth emits compared to an ideal black body,
- $\sigma = 5.67 \cdot 10^{-8} \frac{\text{W}}{\text{m}^2\text{K}^4}$ is the Boltzmann constant,
- T is the mean temperature on the earth's surface measured in Kelvin,
- $4\pi R^2$ is the area of the entire earth's surface (because radiation takes place in all directions as shown in the following figure).

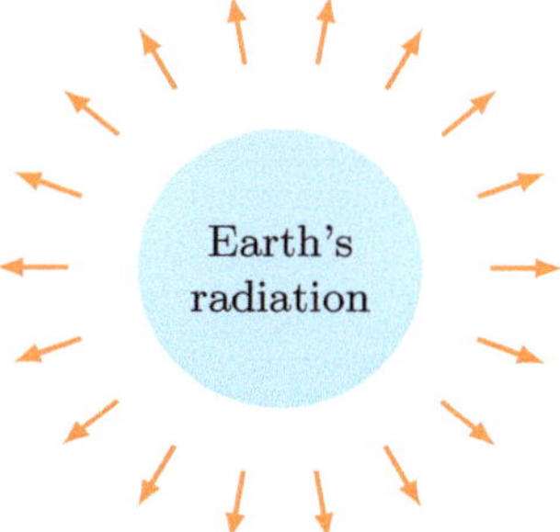

The decisive **radiation factor** $g < 1$, which can be influenced, models the greenhouse gas effect. It reflects the fact that the emission temperature in the real atmosphere – that is, the temperature that one would see from space – does not correspond to the temperature on the earth's surface.

The reason is that a large portion of the infrared radiation emitted by the earth's surface does not leave the atmosphere because it is absorbed by greenhouse gases, mainly water vapor and CO_2, that are present in the atmosphere. These particles radiate the absorbed energy in all directions, including downwards back to the earth. As a result, the earth's surface heats up and, by convection, the entire atmosphere heats up as well. Due to this **greenhouse effect**, the average temperature on the earth's surface is higher than without an atmosphere.

In the convective atmosphere, the temperature decreases with increasing altitude, mainly because the air expands adiabatically with decreasing pressure.

Due to the infrared radiation back into space, which depends on the altitude, the earth system can maintain the balance between incoming and outgoing radiation even in the presence of greenhouse gases. With the current concentration of greenhouse gases, the infrared radiation into space takes place at an altitude of about 5 km.

---

[2] Josef Stefan (1835–1893) was an Austrian mathematician and physicist with Slovenian roots, Ludwig Boltzmann (1844–1906) was an Austrian physicist.

In our present simple energy-balance model we consider the heat loss via infrared emission with a radiation factor of $g < 1$.

Note that its exact value depends on many feedbacks in the earth system. For instance, an increase in the amount of water vapor in the atmosphere reduces the radiation factor g.

The **instantaneous rate of change in the earth's heat energy** per unit of time (physically a power) follows from Equation (2.1):

$$\frac{\mathrm{d}Q}{\mathrm{d}t} = C \cdot M \cdot \frac{\mathrm{d}T}{\mathrm{d}t}.$$

The mass M of the layer of the earth's surface, which is involved in the heat exchange, is given by

$$M = 4\pi R^2 \cdot \Delta R \cdot \rho,$$

where ΔR is the surface layer's mean thickness and ρ its average density.

With the heat equation (2.2), this results in

$$4\pi R^2 \cdot \Delta R \cdot \rho \cdot C \cdot \frac{\mathrm{d}T}{\mathrm{d}t} = P_{\mathrm{gain}} - P_{\mathrm{loss}}.$$

Dividing the equation by $4\pi R^2$ yields the first-order **nonlinear differential equation** for the unknown function $T(t)$, also found in [24], Section 3.2.1:[3]

$$C_E \cdot \frac{\mathrm{d}T}{\mathrm{d}t} = \frac{S_0}{4}(1 - \alpha) - \varepsilon \cdot \sigma \cdot g \cdot T^4.$$

The constant $C_E = \Delta R \cdot \rho \cdot C$ can be interpreted as the *average specific heat capacity per square meter*, because its dimension is $[C_E] = \mathrm{J}/(\mathrm{K} \cdot \mathrm{m}^2)$.

A first-order differential equation not only involves the unknown function $T(t)$, but also its first derivative $\frac{\mathrm{d}T}{\mathrm{d}t}$. The task consists in finding a solution that is a function $T(t)$ (not a number).

At this point, let us refer to the Appendix where the concept of differential equations is introduced.

According to the climate science, the prevailing solar radiation power $\frac{S_0}{4}(1 - \alpha)$ increases by the so-called **radiative forcing** due to feedbacks in the atmosphere, mainly caused by carbon dioxide is aproximately

$$\Delta F = 5.35\frac{\mathrm{W}}{\mathrm{m}^2} \cdot \ln\left(\frac{c}{c_0}\right).$$

Here $c_0 = 280\,\mathrm{ppm}$ [4] is the relative pre-industrial CO_2 reference pollution and c the current relative pollution of CO_2.

---

[3] Reference provided by Prof. Dr. Stefan Brönnimann, Climatology Group at the Geographical Institute of the University of Bern.

[4] ppm is the abbreviation for parts per million, $1\,\mathrm{ppm} = 10^{-6} = 1/10^6 = 1$ millionth, comparable to 1 percent $= 10^{-2} = 1$ hundredth and 1 per mille $= 10^{-3} = 1$ thousandth.

Remark: The contribution of radiative forcing of the other greenhouse gases such as methane, nitrous oxide and others is small in comparance with the effect of CO_2, because aerosols have a negative, cooling effect.

Hence the nonlinear differential equation with radiative forcing is given by

$$C_E \cdot \frac{\mathrm{d}T}{\mathrm{d}t} = \frac{S_0}{4}(1 - \alpha) + \Delta F - \varepsilon \cdot \sigma \cdot g \cdot T^4.$$ (2.3)

The following **Keeling curve**[5] taken from [30] shows the **pure** CO_2 concentration c between the years 1958 and 2022:

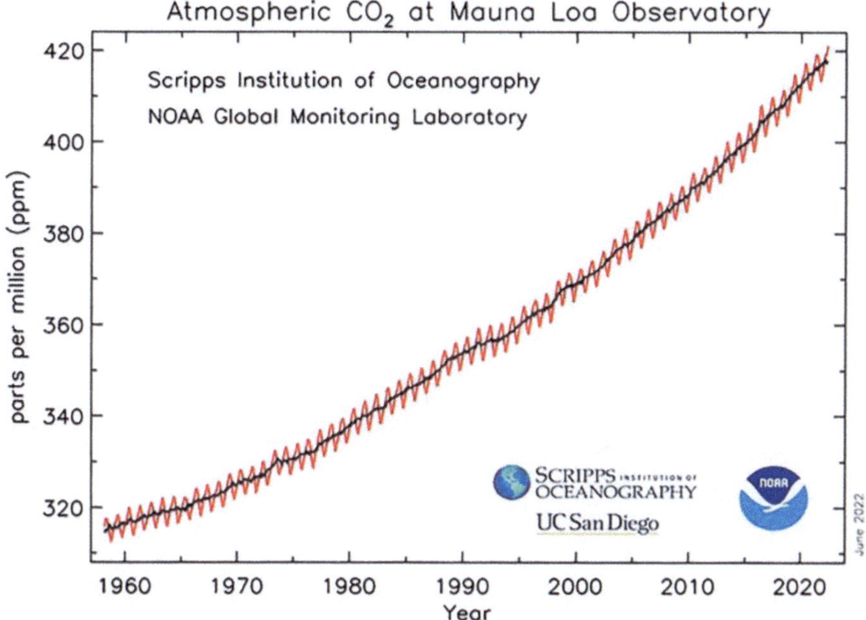

During the short period of 64 years, its value has increased by about 30%! The text for the diagram goes as follows:

In 2024, the global average annual CO_2 concentration reached a record high of 425 ppm. This value compares to the peak value in the Pliocene 4.1 to 4.5 million years ago!
According to the Keeling curve, c as a function of time has been almost linearly increasing since 1999.
The quantity ΔF depends logarithmically on c and therefore changes only slightly. Let us, in a first approximation, consider ΔF to be constant.

---

[5] Charles David Keeling (1928–2005) was a US climate researcher and a professor of chemistry at the Scripps Institution of Oceanography near San Diego. He held visiting professorships at the Universities of Heidelberg (1969–1970) and Bern (1979–1980).

That is, we equate the right-hand side of the differential equation (2.3) to zero. Using the numerically given parameters that depend on the radiation factor g, we obtain the following time-constant solution in Kelvin:

$$T_\infty(g) = \sqrt[4]{\frac{S_0(1-\alpha)/4 + \Delta F}{\varepsilon \cdot \sigma \cdot g}} = \sqrt[4]{\frac{1370 \cdot 0.68/4 + \Delta F}{0.97 \cdot 5.67 \cdot 10^{-8} g}} = \sqrt[4]{\frac{233 + \Delta F}{5.500}} \frac{100}{\sqrt[4]{g}} \tag{2.4}$$

In physics this is called a **stationary solution**.

In 2024 there were 425 ppm CO_2 in the atmosphere:

$$\Delta F = 5.35 \frac{\mathrm{W}}{\mathrm{m}^2} \cdot \ln\left(\frac{425}{280}\right) = 2.23 \frac{\mathrm{W}}{\mathrm{m}^2}.$$

This value is small compared to the power $\frac{S_0}{4}(1-\alpha)\,\mathrm{W/m^2}$, which justifies Equation (2.4):

$$T_\infty(g) = \frac{256}{\sqrt[4]{g}} \quad \text{in Kelvin.} \tag{2.5}$$

The following diagram taken from [74] shows the so-called **temperature anomalies** in degree Celsius.[6]

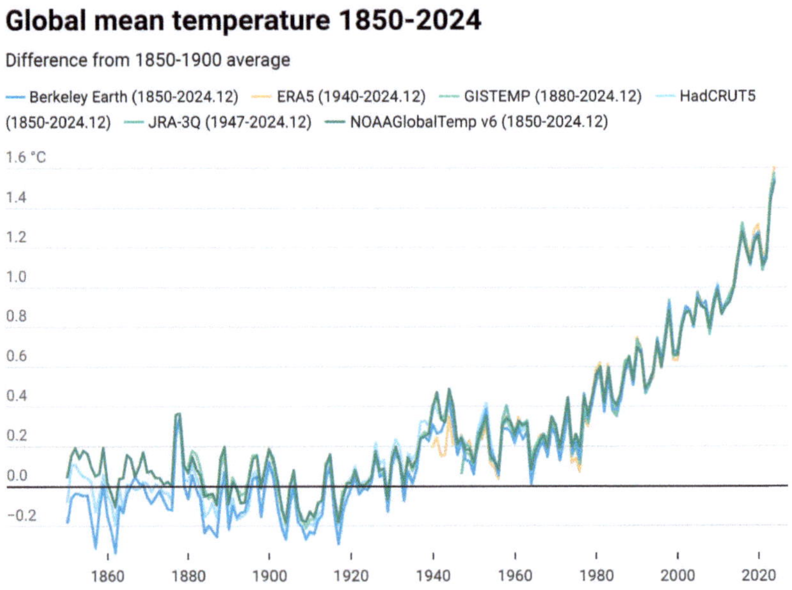

Global mean temperature 1850-2024

Difference from 1850-1900 average

Annual global mean temperature anomalies relative to a pre-industrial (1850–1900) baseline shown from 1850 to 2024

Chart: WMO · Created with Datawrapper

---

[6] Anders Celsius (1701-1744), professor of Astronomy at Uppsala University in Sweden, originally defined his temperature scale backward: $0\,°\mathrm{C}$ for the boiling point of water and $100\,°\mathrm{C}$ for the freezing point of water.

Also from [74]: „The clear signs of human-induced climate change reached new heights in 2024, which was likely the first calendar year to be more than 1.5 °C above the pre-industrial era, with a global mean near-surface temperature of 1.55±0.13 °C above the 1850-1900 average. Record greenhouse gas concentrations combined with El Niño and other factors drove 2024 to a record heat: the warmest year in the 175-year observational record."

The trend between the years 1970 and 2020, a period of 50 years, was roughly linear, with an approximately linear increase of $(1.24 - 0.26)/5 = 0.20\,°C$ per decade.

From [71]: „According to the Intergovernmental Panel on Climate Change (IPCC)'s assessment, the decade 2011–2020 was the warmest decade since the last interglacial period about 125,000 years ago and the last six years between 2015 and 2020 were also the warmest years since measurements began. As can be seen from the above diagram, the year 2024 was 1.55 °C above the pre-industrial value (1850–1900) of 13.7 °C."[7]

According to both Copernicus and NASA, the mean global temperature was 15.0 °C in 2023 and 15.2 °C in 2024 with 1.46 °C above the average pre-industrial temperature 1850–1900. Therefore, for the calculation in the following section let the temperature anomaly at the **beginning of 2025** be

$$1.50\,°C \tag{2.6}$$

with the mean surface temperature $273 + 13.7 + 1.5 = 288$ K.
Due to (2.5), **the stationary temperature increase** $\Delta_\infty(g)$ is given by

$$\Delta_\infty = \frac{256}{\sqrt[4]{g}} - 288 \quad \text{in } °C. \tag{2.7}$$

Without greenhouse gases (i.e., $g = 1$), we would have the mean surface temperature $\Delta_\infty(1) = 256 - 288 = -32\,°C$!

Solving Equation (2.7) for the radiation factor g results in

$$g = \left(\frac{256}{\Delta_\infty + 288}\right)^4.$$

Here are some g values for the much-discussed scenarios:

- For $\Delta_\infty = 0\,°C$ the result is $g_0 = 0.624$.
- For $\Delta_\infty = 1.5\,°C$ the result is $g_{1.5} = 0.6115$.
- For $\Delta_\infty = 2.0\,°C$ the result is $g_{2.0} = 0.6073$.
- For $\Delta_\infty = 3.0\,°C$ the result is $g_{3.0} = 0.600$.

Comparing $g_0 > g_{1.5} > g_{2.0} > g_{3.0}$ confirms the fact that the larger g, the smaller the temperature increase Δ_∞.

---

[7] The mean temperature anomaly of the land surface was 2.28° C and that of the water surface of the oceans was 1.15° C, as the land heats up more than the oceans. Source: [59].

A key variable in climate science is **climate sensitivity**, defined as the temperature rise in the event that the c value is double that of the pre-industrial age. The radiative forcing would be $\Delta F = 5.35 \cdot \ln 2 = 3.71 \, \text{W/m}^2$.

Statement of the 2017 IPCC's fifth assessment report: *The climate sensitivity is probably between* $1.5\,^\circ C$ *and* $4.5\,^\circ C$.
The average of $3.0\,^\circ$C corresponds to the fourth scenario mentioned above with $g_{3.0} = 0.600$.

2.2 Temperature increase over time for various scenarios

First we compute the numerical value of $C_E = \Delta R \cdot \rho \cdot C$.

- The penetration depth of light in water depends on the wavelength. It varies between 5 m (for red light) and 60 m (for blue light). But the relevant layer of water is mainly given by the convection depth, which ranges between 10 m and above 100 m. It is called the *mixed layer*. Its average thickness is influenced by the wind (impulse drive) and by the rate of change in water density on the surface caused by warm and fresh water flows (so-called buoyancy forces). The average global mixed layer is estimated at $\Delta R \approx 40$ m.

- Approximately 3/4 of the earth's surface is covered by water with a specific heat capacity of $4187 \frac{\text{J}}{\text{kg·K}}$. Taking into account the fact that the different materials of the earth's crust have smaller heat capacities, the specific heat capacity averaged globally is estimated at $C \approx 3440 \frac{\text{J}}{\text{kg·K}}$.

- The mean density is estimated at $\rho \approx 1500 \, \text{kg/m}^3$.

This results in the specific heat capacity per square meter

$$C_E \approx 40 \text{ m} \cdot 1500 \, \frac{\text{kg}}{\text{m}^3} \cdot 3440 \, \frac{\text{J}}{\text{kg} \cdot \text{K}} = 2.06 \cdot 10^8 \, \frac{\text{J}}{\text{m}^2 \cdot \text{K}}.$$

Because $\text{J} = \text{W} \cdot \text{s}$ (1 joule = 1 watt second), the time t is measured in seconds.

Let $T(t) = T_1 + \Delta(t)$ with $T_1 = 288$ K and $\Delta(t)$ in $^\circ$C. In order to calculate the temperature profile $\Delta(t)$ over time, the **differential equation is linearized** as follows:
Since Δ is small relative to T and using the binomial theorem, the following is a good approximation:

$$T^4 = (T_1 + \Delta)^4 \approx T_1^4 + 4T_1^3 \cdot \Delta,$$

The reason is that the remaining terms containing the factors Δ^2, Δ^3, Δ^4 compared to the above right-hand side are so small that they can be neglected.

Since $T(t)$ and $\Delta(t)$ differ only by the additive constant T_1, we have $\frac{\text{d}\Delta}{\text{d}t} = \frac{\text{d}T}{\text{d}t}$.

The linearized differential equation for the function $\Delta(t)$ follows from (2.3):

$$C_E \cdot \frac{\mathrm{d}\Delta}{\mathrm{d}t} = A - B \cdot \Delta \qquad \text{with } \Delta(0) = 0. \tag{2.8}$$

Here $A(g) = \dfrac{S_0}{4}(1-\alpha) + \Delta F - \varepsilon \cdot \sigma \cdot T_1^4 \cdot g$ and $B(g) = 4 \cdot \varepsilon \cdot \sigma \cdot T_1^3 \cdot g.$

Hence (2.8) provides us with a **linear model.**

For the 12-year time span 2025-2036, we use a linear extrapolation of about 2.5 ppm/year for c (see Keeling curve): $\Delta F = 5.35 \cdot \ln(\frac{425+30}{280}) = 2.60 \frac{\mathrm{W}}{\mathrm{m}^2}$ for 2036. The arithmetic mean with the calculated value 2.23 $\frac{\mathrm{W}}{\mathrm{m}^2}$ for 2024 is 2.41 $\frac{\mathrm{W}}{\mathrm{m}^2}$. Considering effects of the non-CO_2 greenhouse gases and the water vapor in the atmosphere, we increase it to 3.0 $\frac{\mathrm{W}}{\mathrm{m}^2}$:

$$\frac{S_0}{4}(1-\alpha)+3.0 = \frac{1370}{4} \cdot 0.68 + 3 = 236 \frac{\mathrm{W}}{\mathrm{m}^2}, \quad \varepsilon \cdot \sigma = 0.97 \cdot 5.67 \cdot 10^{-8} = 5.500 \cdot 10^{-8}.$$

With

$$T_1^4 = 6.8797 \cdot 10^9, \quad T_1^3 = 2.3888 \cdot 10^7,$$

this results in the linear expression of g:

$$A(g) = 236 - 378 \cdot g \quad \text{and} \quad B(g) = 5.255 \cdot g. \tag{2.9}$$

We obtain the limit Δ_∞ of $\Delta(t)$ for $t \to \infty$ as a function of g by equating $\frac{\mathrm{d}\Delta}{\mathrm{d}t} = 0$. That is, by equating to zero the right-hand side of (2.8):

$$\Delta_\infty(g) = \frac{A(g)}{B(g)} = \frac{236 - 378 \cdot g}{5.255 \cdot g} \quad \text{in } {}^\circ\mathrm{C}. \tag{2.10}$$

Here we can check the case $\Delta_\infty = 0$. $A(g) = 0$ must hold. This again results in $g = g_0 = 0.624$. The other values calculated in this way also agree with earlier calculations.

The **Intergovernmental Panel on Climate Change (IPCC)** in [10] contains extensive information, including the following:
"Each of the past four decades since 1850 has been warmer than any previous decade.
Compared to the period 1850–1900, the mean global surface temperature increased

- in the first two decades of the 21st century from 2001–2020 by 0.99 [0.84 to 1.10] $^\circ$C;
- in the period 2011–2020 by 1.09 [0.95 to 1.20] $^\circ$C; over land by 1.59 [1.34 to 1.83] $^\circ$C, over the sea by 0.88 [0.68 to 1.01] $^\circ$C;
- in 2021 by 1.21 $^\circ$C."[8]

---

[8] According to the UN's World Meteorological Organization (WMO), the global increase in average temperature in 2021 was 1.11 $^\circ$C above the pre-industrial level. This statement differs from the IPCC's of 1.21 $^\circ$C by 0.1 $^\circ$C.

The temperature increases of $2.0\,°\mathrm{C}$ or $1.5\,°\mathrm{C}$, respectively, proposed by the Intergovernmental Panel on Climate Change, refer to the earth's **pre-industrial mean surface temperature from 1850 to 1900.**

We consider, as a third case, a temperature increase of $3.0\,°\mathrm{C}$ compared to the pre-industrial mean surface temperature.

$A(g)$ and $B(g)$ in the table are computed by use of Equation (2.9):

g value	$A(g)$	$B(g)$	$A(g)/B(g)$
$g_{1.5} = 0.6115$	4.91	3.213	1.53
$g_{2.0} = 0.6073$	6.50	3.191	2.03
$g_{3.0} = 0.600$	9.26	3.153	2.94

The last column is for checking.

Equation (2.8) implies the linear differential equation

$$\frac{\mathrm{d}\Delta}{\mathrm{d}t} = \frac{A(g)}{C_E} - \frac{B(g)}{C_E} \cdot \Delta \qquad \text{with } \Delta(0) = 0.$$

Time $t = 0$ refers to the year 2025.

Assuming that the g value is constant over time, the solution to the initial value problem for $\Delta_\infty(g) \neq A(g)/B(g)$, and with (2.10) in mind, is

$$\Delta(t) = \Delta_\infty(g) \cdot \left[1 - \exp\left(-\frac{B(g)}{C_E} \cdot t \right) \right] \qquad \text{with } t \text{ in seconds} \qquad (2.11)$$

and for $\Delta_\infty(g) = A(g)/B(g)$, it is the stationary solution $\Delta(t) = \Delta_\infty(g)$ (see Example 9.2 in the Appendix).

Conversion from t seconds to τ years: $1 \text{ year} = 365 \cdot 24 \cdot 3600 \text{ s} = 3.15 \cdot 10^7 \text{ s}$. With the new constant in the exponent

$$-B(g) \cdot \frac{3.15 \cdot 10^7}{2.06 \cdot 10^8} = -0.1529 \cdot B(g)$$

the result is

$$\Delta(\tau) = \Delta_\infty(g) \cdot (1 - e^{-0.1529 \cdot B(g) \cdot \tau}) \qquad \text{with } \tau \text{ in years.} \qquad (2.12)$$

Because of (2.6), the difference of $1.50\,°\mathrm{C}$ has already been used up at the beginning of 2025. Therefore, we have to reduce the factor $\Delta_\infty(g)$ in (2.12) by $1.50\,°\mathrm{C}$ and add $1.50\,°\mathrm{C}$ at the end of $\Delta(\tau)$ to obtain the solutions for temperature increases of 1.5, 2.0 and $3.0\,°\mathrm{C}$ relative to pre-industrialization:

$$\Delta_{1.5}(\tau) = 1.50 \quad \text{constant,}$$

$$\Delta_{2.0}(\tau) = 0.50 \cdot (1 - e^{-0.4879 \cdot \tau}) + 1.50,$$

$$\Delta_{3.0}(\tau) = 1.50 \cdot (1 - e^{-0.4821 \cdot \tau}) + 1.50.$$

Tempeature increases over time:

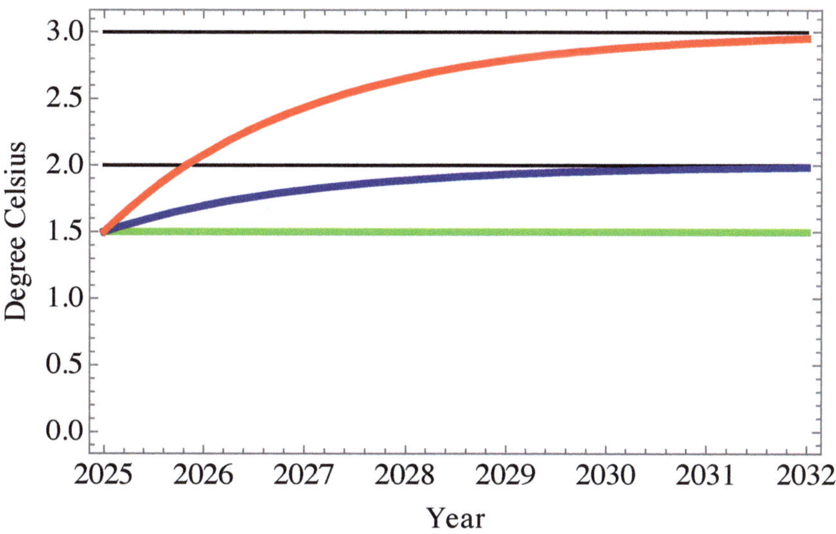

Remark: In a time range of approximately 20 years the calculated constant C_E is reasonable. For longer periods of time, up to decades and centuries, one must consider that the seas warm up to much greater depths, i.e. the quantity of ΔR would be a multiple of 40 m.

2.3 The warmest year on record was 2024

The World Meterological Organisation WMO wrote on 10 January 2025:

„WMO confirms 2024 as warmest year on record at about 1.55 °C above pre-industrial level.

Key messages:

- The past ten years 2015–2024 are the ten warmest years on record.
- Six international datasets are used to reach the consolidated WMO global figure.
- 2024 saw exceptional land and sea surface temperatures and ocean heat.
- Long-term temperature goal of the Paris Agreement not yet dead but in grave danger.

The global average surface temperature was 1.55 °C (with a margin of uncertainty of ±0.13 °C) above the 1850–1900 average, according to WMO's consolidated analysis of the six datasets. This means that we have likely just experienced the first calendar year with a global mean temperature of more than 1.5 °C above the 1850–1900 average.

In September 2024, the global mean temperature anomaly was $1.54\,°C$ above the pre-industrial level and it was the 14th month in a 15-month period for which the global-average surface air temperature exceeded $1.5\,°C$.[9]

The **global mean surface temperature in 2024 was**
$(273.1 + 13.7 + 1.54)\,\text{K} = 288.2\,\text{K}$."

Researchers led by Jens Daniel Müller and Nicolas Gruber (ETH Zürch) reported that in **2023, the global ocean absorbed approximately 1 Gt less CO_2** than expected – equivalent to a **10% decline** in uptake compared to previous years. This shortfall corresponds to about **half of the EU's annual CO_2 emissions.** The main cause is record high sea surface temperatures, especially in the North Atlantic.[10]

The mean stationary (limit) temperatures of the ocean, the land and the entire earth's surface are calculated in [17], using the zero-dimensional energy balance model for the mean stationary temperature rises of $0\,°C$, $1.5\,°C$, $2.0\,°C$, $3.0\,°C$ **since the beginning of 2025.**
By adding (2.6) we obtain the mean stationary temperatures **since the pre-industrial time,** corresponding to $1.5\,°C$, $3.0\,°C$, $3.5\,°C$, $4.5\,°C$.
Here is a graphical representation:

Stationary temperatures for ocean, land and global surface

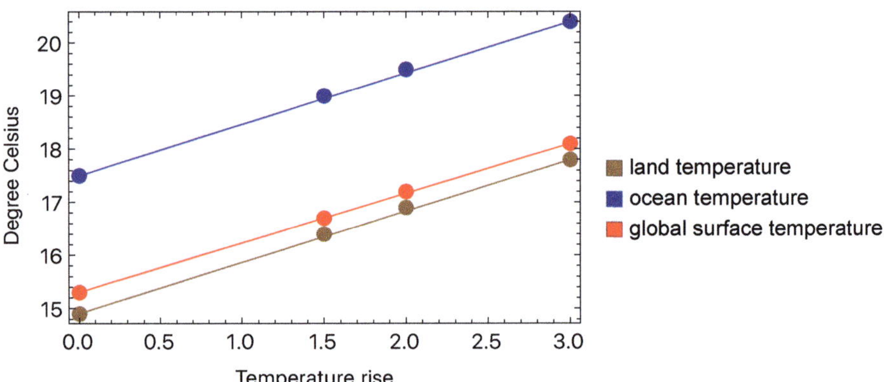

An increase of the temperature rise by $1.00\,°C$ leads to an increase of the global mean surface temperature of $1.00\,°C$. The three lines are approximately parallel with slope 1.00.
The mean temperature of the ocean surface is always $2.6\,°C$ higher than that of the land surface.

---

[9] see https://climate.copernicus.eu/surface-air-temperature-september-2024.

[10] Source: Müller, J. D. et al.: Unexpected decline in the ocean carbon sink under record-high sea surface temperatures in 2023, *Nature Climate Change (2025).*

Chapter 3
Increase in greenhouse gases

3.1 Human polluter

When looking at the Keeling curve in Section 2.1, following Equation (2.3), the central question arises as to whether the increase in the proportion of CO_2 in the atmosphere over the past few decades was caused by humans. The answer is unmistakably: Yes! Man is the culprit.[1]

Reason:
The following diagram, taken from [64], shows measurements of the concentration of the carbon isotope C14 in the atmospheric CO_2 from 1945 to 2015. The concentration in the atmosphere before 1950, caused by cosmic rays, was practically constant for thousands of years. C14 carbon dating is based on this fact.[2]

The above-ground atomic bomb tests in the 1950s and 1960s increased the concentration of C14 for short periods of time (bomb peak).

Intensive burning of fossil fuels does **not** produce C14, solely C12. As a result, the concentration of C14 in the atmosphere after the bomb peak was continually diluted so that in 2015 it fell practically back to the level before the atomic bomb tests.

An analysis in [24] shows significantly smaller concentrations of C14 than in the pre-industrial age for various scenarios up until the end of this century.

---

[1] From the *Neue Zürcher Zeitung* (NZZ) Planet A newsletter of April 10, 2023: The outgoing President of the World Bank, David Malpass, was asked at a panel discussion in September 2022 whether the burning of fossil fuels is causing man-made climate change. Instead of saying yes, he evaded by answering that he was not a scientist.

[2] See Sections 3.3.2 and 3.3.3 in [16].

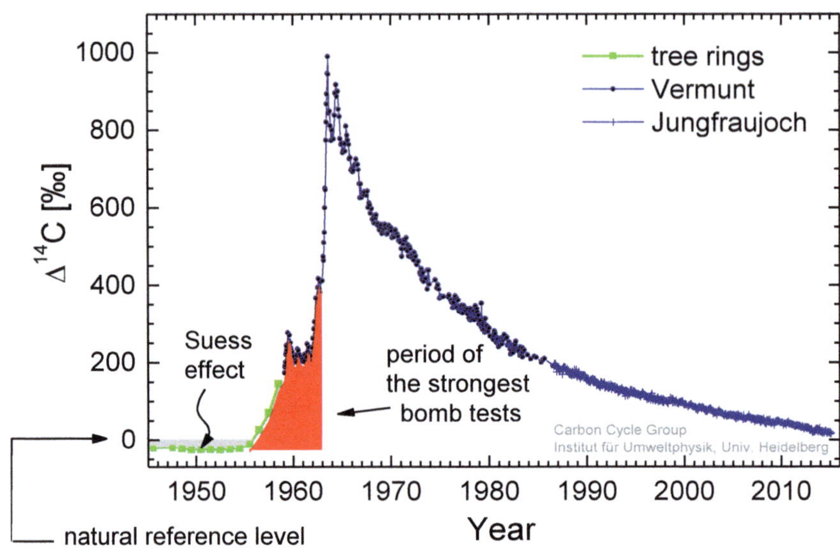

Here the definition is

$$\Delta^{14}\mathrm{C} = \left(\mathrm{f} \cdot \frac{\mathrm{n}^{14}}{\mathrm{n}^{C}} - 1000 \right) ‰, \tag{3.1}$$

where $f = 8.19 \cdot 10^{14}$ is a dimensionless constant, n^{14} designates the number of C14 atoms and n^{C} the number of CO_2 molecules in a given volume of the atmosphere.

Jungfraujoch in Switzerland has an elevation of 3450 m above sea level, Vermunt in Austria that of 2037 m.

- For the practically constant C14 concentration $r = \frac{n^{14}}{n^C} = 1.22 \cdot 10^{-12}$, which was valid for thousands of years before 1950, we have $\Delta^{14}\mathrm{C} = 0‰$.
- Double the C14 concentration $2r = 2.44 \cdot 10^{-12}$ yields $\Delta^{14}\mathrm{C} = 1000‰$.
- 1.5 times the C14 concentration $1.5r = 1.83 \cdot 10^{-12}$ yields $\Delta^{14}\mathrm{C} = 500‰$.

From other sources, we know that $\Delta^{14}\mathrm{C} \approx 190‰$ in 1987 and $\Delta^{14}\mathrm{C} \approx 20‰$ in 2017. It is remarkable that these global data practically agree with the values of the diagram above.

Conclusion: There is no doubt that the proportion of CO_2 in the atmosphere is man-made! For more information on this consult [32].

3.2 CO_2 concentration and temperature over hundreds of thousands of years

By analyzing ice cores, the CO_2 concentration in the atmosphere can be traced back hundreds of thousands of years. Here is one of several diagrams taken from [66], illustrating the historical evolution of both the CO_2 concentration and the temperature during the last 800,000 years:

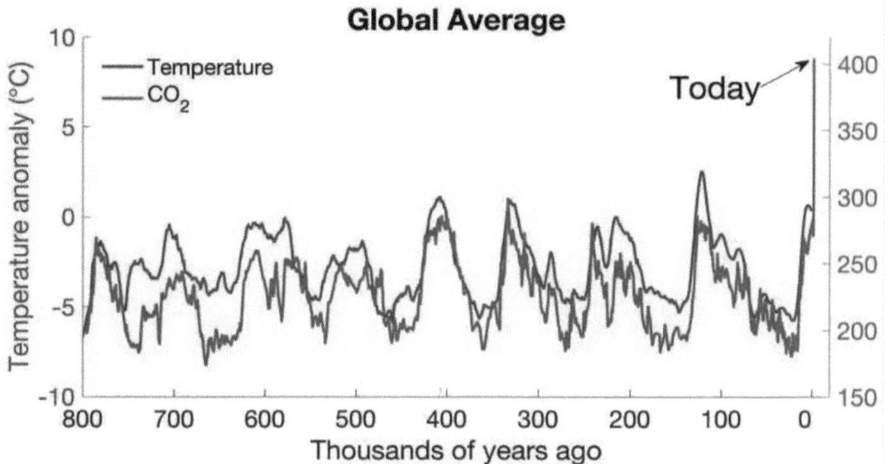

It shows the worrisome one-time increase in the recent past in an extremely short time interval. The region on the right of the peak is zoomed in on the Keeling curve (in Section 2.1).

The argument that there have always been fluctuations is meaningless, since we are dealing with time intervals of the order of 100,000 years, in contrast to short time periods of just a few decades!

Statement by Prof. Dr. Reto Knutti, ETH Zurich, former member of the Intergovernmental Panel on Climate Change IPCC:[3]

"Climate change is developing into a climate crisis. Since the industrialization, mankind has increased the CO_2 concentration in the air by 50 percent. Comparable values go back millions of years, long before humans existed."

Excerpt from a press release in January 2025 of www.beyondepica.eu/en/:

Antarctica: Historic Drilling Campaign Reaches more than 1.2-Million-Year-Old Ice.

At the remote Little Dome C site in Antarctica, a research team representing twelve scientific institutions from ten European nations has just achieved a historic milestone for climate science. As part of the European-funded *Beyond EPICA - Oldest Ice project*, the team successfully concluded a decisive drilling

---

[3] From *Neue Zürcher Zeitung*, April 23, 2021.

campaign, reaching the depth of 2,800 meters, where the Antarctic ice sheet meets the bedrock.

The extracted ice preserves an unprecedented record of Earth's climate history, continuous information on atmospheric temperatures and pristine samples of old air with greenhouse gases spanning over 1.2-million-year-old ice and probably beyond.

The European teams in the field have accomplished an impressive achievement: a total of more than 200 days of successful drilling and ice core processing operations across four field seasons in the harsh environment of the central Antarctic plateau at an altitude of 3,200 meters above sea level and with an average summer temperature of -35 °C.

The ice core will offer unprecedented insights into the Mid-Pleistocene Transition, a remarkable period between 900,000 and 1.2 million years ago when glacial cycles slowed down from 41,000-year to 100,000-year intervals. The reasons behind this shift remain one of climate science's enduring mysteries, which this project aims to unravel.

The ice samples will be analysed by laboratories in Germany, Switzerland, Italy, France and Great Britain. The University of Bern will use an *innovative laser beam technique*, that it has developed together with Swiss Federal Laboratories for Materials Science and Technology (EMPA).[4] This allows measuring the greenhouse gases in the ice with extreme precision without contaminating the samples with ambient air and without melting the ice. All that is needed is an ice sample one centimeter in thickness.

---

[4] Prof. Hubertus Fischer, successor of Prof. Thomas Stocker at the University of Bern, is the Swiss Principal Investigator of the EU Project *Beyond EPICA - Oldest Ice*.

3.3 Paleoclimatology

In palaeoclimatology, the climate over the course of the earth's history is analyzed on the basis of geological facts and model calculations. So-called *proxy data* are needed to reconstruct early climate conditions.

Examples of proxy data:

1. *Historical documents* contain comprehensive information about the climate in the past.
2. *Corals* build their hard skeleton from calcium carbonate, a mineral found in seawater. The carbonate contains oxygen isotopes and trace materials that are used to determine the temperature of the sea in which the coral grew.
3. *Pollen*: All flowering plants produce pollen grains. Their different shapes can be used to identify the type of plant. Because pollen grains are well preserved in the lake or sea floor, analyzing each layer helps clarifying which plants were present at the time the sediment was deposited. Based on the plant types, conclusions can be drawn about the climate of the region in question.
4. *Ice cores*: They contain inclusions of dust, air bubbles and oxygen isotopes, depending on the particular environment. Such data reflect the climate over hundreds of thousands of years.
5. *Tree rings*: Climatic conditions affect their spacing, the density of the wood and the composition of the various isotopes. Such data reflect the climate over thousands of years.
6. *Sediments in seas and lakes*: Large amounts of sediments on the bottom of seas and lakes provide information about the region in the past. Cores drilled into the sediment layer of the sea or lake floor contain tiny fossils and chemicals and yield additional interpretations of the past climate.

Analyzing proxy data gives us a broader understanding of the climate well beyond the period of recordings by means of instruments.

A scientifically prominent region is located around *Crawford Lake*, which is only 200 m in diameter. It is near Toronto, Ontario, Canada.

- Analyzing pollen allows reconstructing the history of this region over several hundred years.
- Geochemical investigations of the sediment cores enable scientists to reconstruct the atmospheric CO_2 pollution over a period of about 150 years.

The *Anthropocene* describes the most recent geological epoch brought about by man. In 2023, its beginning was set to 1950. Reference: Cores from Crawford Lake sediments.

The following diagram with various future scenarios is taken from the journal *Science* [56]:

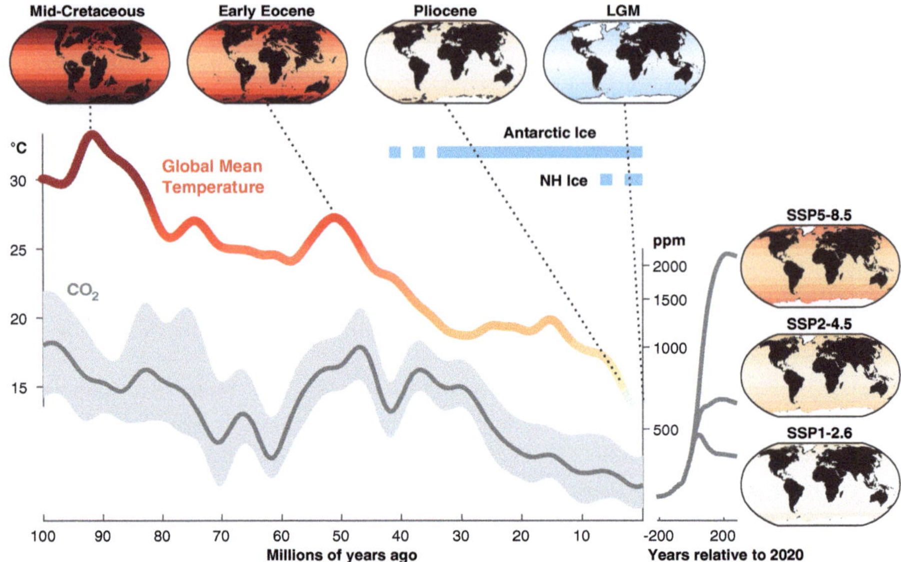

It shows both the global mean temperature of the earth's surface and the CO_2 concentration over the past 100 million years.

The horizontal blue lines mark the period of time with massive ice masses. LGM on the upper right stands for Last Glacier Maximum, NH for Northern Hemisphere, SSP on the right stands for Shared Socioeconomic Pathway.

Earlier climates were very different from today's. Global temperatures, extent of polar ice, regions of deep-water formations, types of vegetation, patterns of precipitation, evaporation and their variability were quite diverse over large periods of time. Such differences provide invaluable information because they provide evidence of how in climate processes over certain regions CO_2 concentrations of the past can be associated with those of the future.

The curves on the right belong to various scenarios for the next 200 years:

- Under the sustainable SSP1-2.6 scenario, in which emissions are reduced with a negative balance by the end of the 21st century, the CO_2 concentration roughly corresponds to that of the Pliocene[5] period.
- The fossil-intensive SSP5-8.5 scenario would lead the CO_2 concentration to values of the *Cretaceous period*, which began about 145 million years ago and ended about 66 million years ago.

The diagram shows that the current CO_2 pollution is lower than in the past 100 million years. More on this in [42].

---

[5] The Pliocene period was an era from 1.5 million years ago to 5 million years ago.

Chapter 4
Carbon budget for various scenarios

The atmosphere does not decompose the CO_2 emissions. It accumulates the emissions, like a liquid container into which water is repeatedly poured.

The idea of the mathematical model used comes from various publications, including [29]. The model was derived from this publication by Reto Knutti (ETH) and then further developed by Nicolas Gruber (ETH).

4.1 Goal and model

This section deals with maintaining a specified level for the global mean temperature increase

- by calculating the maximum permissible amount of carbon emitted into the atmosphere, the so-called carbon budget;
- by determining the time span until the carbon budget is used up.

To do this, we use a simple model that allows calculating the global carbon budget:

$$\Delta W = \Delta F - \Delta R \qquad \text{(power per square meter).}$$

- ΔW denotes the amount of net power absorbed by the earth (mainly by the oceans) in $\frac{W}{m^2}$.
- $\Delta F = f \cdot \ln(\frac{c}{c_0})$, where $f = 5.35 \frac{W}{m^2}$, denotes the radiative forcing caused by the increase in CO_2 concentration.
- ΔR denotes the power in $\frac{W}{m^2}$ reflected into the atmosphere.

In a first approximation, ΔW and ΔR are proportional to the temperature increase ΔT:

$$\Delta W = \kappa \cdot \Delta T, \qquad \Delta R = \lambda \cdot \Delta T.$$

A. Fässler, *Mankind's Problem: Climate Change*,
https://doi.org/10.1007/978-3-662-71846-9_4

An analysis of the observed warming, taking into account the heat absorption of the oceans and using climate models, provides the parameters

$$\kappa = 0.6 \pm 0.05 \, \frac{W}{m^2 K} \text{ and } \lambda = 1.4 \pm 0.05 \, \frac{W}{m^2 K}.$$

4.2 Budget for 1.5°C limit and carbon data

The first step is to calculate the CO_2 concentration $c = c_{1.5}$ for the internationally agreed temperature increase of $\Delta T = 1.5\,°C$:

$$\Delta F = \Delta W + \Delta R = f \cdot \ln\left(\frac{c_e}{c_0}\right) = (\kappa + \lambda) \cdot \Delta T.$$

Assuming that the non-CO_2 radiative forcings[1] are responsible for about 10% of the temperature increase of $1.5\,°C$, the value $\Delta T = 1.35\,°C$ must be used in the calculation:

$$c_{1.5} = c_0 \cdot \exp\left(\frac{(\kappa + \lambda) \cdot 1.35}{f}\right) \text{ ppm.} \tag{4.1}$$

Model calculations show significant percentage deviations. For example, the sum of parameters $\lambda + \kappa = 2.0 \pm 0.07$ appearing in the exponent delivers

$$c_{1.5} = 464 \pm 8 \text{ ppm.} \tag{4.2}$$

Compared to the mass of the carbon atom C, the mass of the CO_2 molecule is greater by a factor of

$$q = \frac{2 \cdot 16 + 12}{12} = 3.67,$$

because the relative atomic masses of the oxygen atom O and the carbon atom C are 16 and 12, respectively.

In the following, 1 Gt stands for 1 giga ton = 1 billion tons = 10^9 tons.
It is known that an increment of the c value by 1 ppm increases the pure carbon mass M in the atmosphere by 2.13 GtC. [2]

Total mass of carbon in the atmosphere by the end of 2024 with 420 ppm was 895 GtC.

---

[1] This involves other harmful substances, primarily methane CH_4 (the portion caused by cows is about 30%, that caused by agriculture is about 45%) and nitrous oxide N_2O. It turns out that the same amount of methane is about 24 times, and nitrous oxide is 398 times more harmful than CO_2. However, they are found in the atmosphere in significantly lower concentrations than CO_2.

[2] Deduction: Total mass of the atmosphere is $\approx 5.15 \cdot 10^6$ Gt. Dry air contains approximately 78% nitrogen N, 21% oxygen O, and 1% argon Ar. The molar masses of N_2, O_2, and Ar are $2 \cdot 14 = 28$ g, $2 \cdot 16 = 32$ g, and 40 g, respectively. The average molar mass of an air particle is therefore $0.78 \cdot 28 + 0.21 \cdot 32 + 0.01 \cdot 40 = 29$ g. The molar mass of the molecule CO_2 is $2 \cdot 16 + 12 = 44$ g. And 1 ppm $= 10^{-6}$ of the atmosphere has a mass of 5.15 Gt. The total portion of mass of pure carbon C from CO_2 in the atmosphere is, therefore, 5.15 Gt $\cdot 12/29 = 2.13$ GtC per ppm.

On average, carbon sinks in the oceans and in the forests remove about half of the emitted mass of M_{em}. This means that only about half of the emissions remains in the atmosphere, a ratio that is known as the so-called **airborne fraction** f_{air}. It is defined as the ratio of the total emitted pure carbon mass M_{em} and the carbon mass M remaining in the atmosphere:

$$f_{air} = \frac{M}{M_{em}}$$

This means for $f_{air} = 50\%$ that 50% of M_{em} stays in the atmosphere and 50% is absorbed by the oceans and the biosphere.

More information can be found in *Nature Communications 15, No.8507, October 2024*: „Empirical analysis leads to the estimate for the airborne fraction of $(47 \pm 1.1)\%$ over the period 1959–2022."

However, model simulations show that the overall trend for f_{air} in the future increases with the rising temperature $T(t)$.[3]

Therefore, we use 0.50 instead of 0.47 for f_{air} in the following calculation.

Thus by increasing the atmospheric CO_2 concentration from c_0 to $c_{1.5}$, the total mass $M_{em} = \frac{M}{f_{air}}$ of pure carbon in the atmosphere increases to

$$M_{em} = \frac{a \cdot (c_{1.5} - c_0)}{f_{air}} = \frac{2.13 \text{ GtC/ppm} \cdot (c_{1.5} - c_0) \text{ ppm}}{0.50} \text{ GtC.} \qquad (4.3)$$

The values (4.2) result in
$$M_{em} = 784 \pm 34 \text{ GtC.} \qquad (4.4)$$

The following diagram taken from [33] shows the CO_2 concentration in parts per million ppm (extended blue Keeling curve) and the annual CO_2 emissions in Gt (black curve):

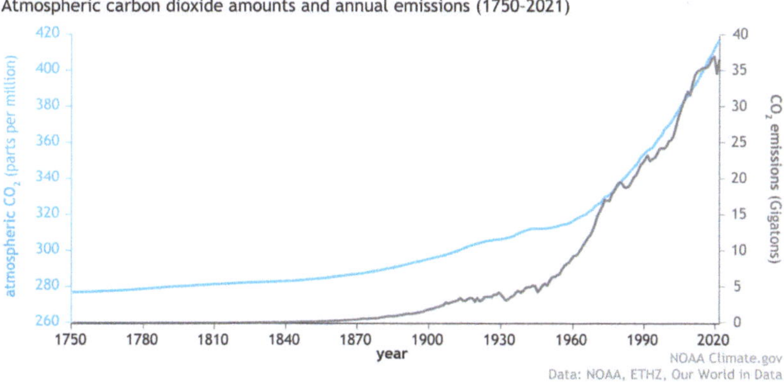

---

[3] More information is given in: *Quantification of the Airborne Fraction of Atmospheric CO_2 Reveals Stability in Global Carbon Sinks Over the Past Six Decades*, Journal of Geophysical Research, Vol.129, Issue 3, 2024. https://agupubs.onlinelibrary.wiley.com.

Here is a graphical representation of the data taken from [55]:

Annual CO_2 emissions in Gt from 2010 to 2024.

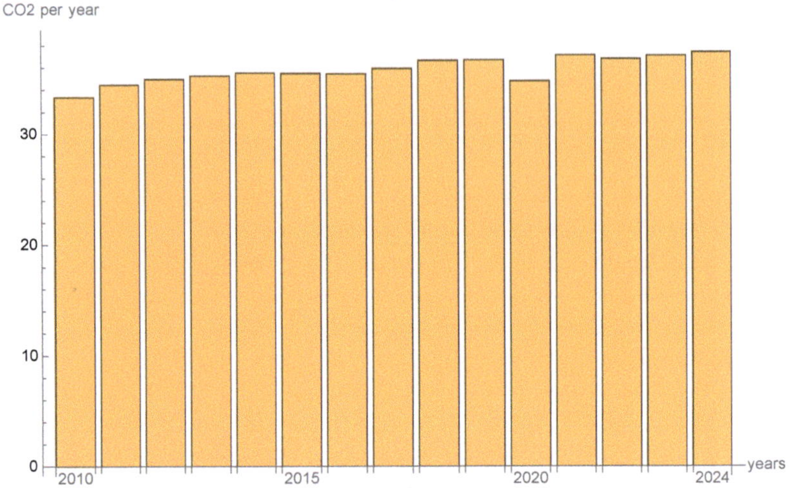

Annual CO_2 emissions from 2018 to 2024:

year	2018	2019	2020	2021	2022	2023	2024
GtCO$_2$	36.7	36.7	**34.8**	37.1	36.8	37.1	37.4

The mass of CO_2 emitted during the COVID year of 2020 only contributed to a reduction of around 6% relative to the previous years 2018 and 2019.

According to the Max Planck Institute in November 2024, increased land use reduces the CO_2 sink, adding another annual 4.2 GtCO$_2$. This results in total emissions of 41.6 GtCO$_2$ for 2024, an increase of 11.2 %.

Total global emissions

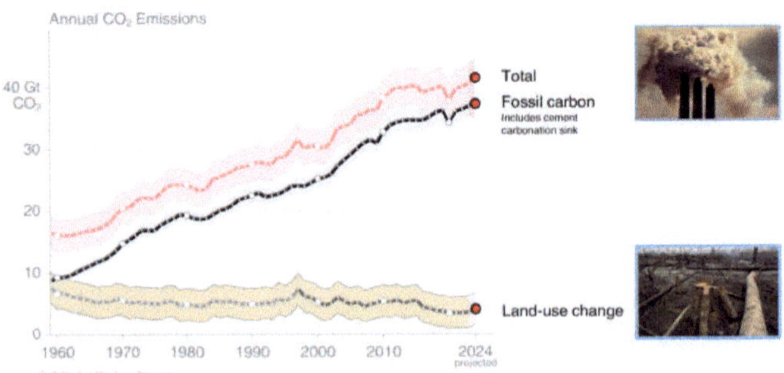

Key sources of land-use-related CO_2 emissions are:

1. *Deforestation*: Cutting down trees reduces the number of carbon sinks. When vegetation is burned or decomposed, the stored carbon is released as CO_2.
2. *Forest degradation*: Logging or damage that doesn't fully remove forests can still emit carbon due to soil disturbance and loss of biomass.
3. *Peatland drainage*: Peatlands store vast amounts of carbon. When they are drained for agriculture or development, they release CO_2 as the organic matter decomposes.
4. *Soil carbon loss from agriculture*: Converting forests or grasslands to cropland or pasture disturbs soil, leading to oxidation of soil organic matter and CO_2 emissions.

From [21], 2025, p.994 and p.1016: „The Cumulative Anthropogenic[4] Emissions CAE (fossil and land use change) for the time span from 1850 to the end of 2024 totalled **CAE = 703 GtC** (2580 $GtCO_2$); that is, 474 GtC from fossil emissions and 229 GtC from land use changes.
For 2023, the emission was 11.1 ± 0.9 GtC. For 2024, it is projected to be 11.4 GtC, 2% above that of 2023.“

From [21], Section 3.5.1 Historical Period 1850–2023, p.994:[5]
„Atmospheric CO_2 concentration was approximately 278 parts per million (ppm) in 1750, reaching 300 ppm in the late 1900s, 350 ppm in the late 1980s, and 419.3 ppm in 2023 (Lan et al., 2024; Fig. 1).
The mass of carbon in the atmosphere increased by 51 % from 590 GtC in 1750 to 890 GtC in 2023. Current CO_2 concentrations in the atmosphere are unprecedented for the last 2 million years, and the current rate of atmospheric CO_2 increase is at least 10 times faster than at any other time during the last 800 000 years (Canadell et al., 2021).“

Because of (4.4), the **budget from January 2025 onwards** is

$$\Delta M_{em} = M_{em} - \text{CAE} = (784 \pm 34)\ \text{GtC} - 703\ \text{GtC} = (81 \pm 34)\ \text{GtC}.$$

The value of **81 GtC** is quite higher than the budget of **35 GtC,** given in the following table of June 2025, but closer to the budget of **65 GtC** in [21] under the title *Executive Summary* with the following text:
„The remaining carbon budget for a 50% likelihood to limit global warming to 1.5, 1.7 and 2.0 °C above the 1850–1900 level has been reduced to 65 GtC (235 $GtCO_2$), 160 GtC (585 $GtCO_2$), and 305 GtC (1110 $GtCO_2$), respectively, from the beginning of 2025, equivalent to around 6, 14, and 27 years, assuming 2024 emission levels.“

The parameters on which the calculation is based are dimensioned in such a way that the avoidance probability of an effective temperature increase below the limit temperature increase is around 50%.

---

[4] caused by human activities, man-made.

[5] Hint from Prof. Guido Grosse, Head of Permafrost Research Unit, Alfred Wegener Institute AWI Potsdam.

If in the near future, humans would cause an annual emission of 11.4 GtC
(as projected in 2024), then from 2025 on, there would be 3 years left until
the budget of 35 GtC is used up by the end of 2027.
But **net zero emissions would have to be achieved abruptly after
3 years, which would be an unrealistic situation!**

Generally one can say, that the mathematical model yields results with a
greater variety, caused by estimations of the parameters. For example: λ and
μ as exponents in the calculation for concentrations c have a strong impact.

From [21] p.1011: „We emphasise the large uncertainty, particularly when
close to the global warming limit of 1.5 °C.“

4.3 Budgets for several scenarios

Let us consider the **temperature increase with limit** $2\,°\mathbf{C}$. Assuming that
the non-CO_2 radiative forcings are responsible for about 15% of the temperature increase of $2\,°C$, then the value $\Delta T = 1.70\,°C$ must be substituted in
the calculation (4.1):

$$c_{2.0} = c_0 \cdot \exp\left(\frac{(\kappa + \lambda) \cdot 1.70}{f}\right). \tag{4.5}$$

Avoidance probablity 50%: $\kappa + \lambda = (2.0 \pm 0.7)\frac{\text{W}}{\text{m}^2\text{K}}$ with $f = 5.35$ yields
$c_{2.0} = (529 \pm 12)$ ppm. A similar calculation using (4.3) with $f_{air} = 0.55$
results in $M_{\text{em}} = (964 \pm 47)$ GtC with the budget
$$\Delta M_{\text{em}} = (964 \pm 47)\text{ GtC} - 703\text{ GtC} = (261 \pm 47)\text{ GtC},$$
comparable to the value 286 GtC shown in the table below.

In the case of a higher **avoidance probability of 67%**, one should use the
smaller sum $\kappa + \lambda = 1.9 \pm 0.7\frac{\text{W}}{\text{m}^2\text{K}}$ in (4.5), resulting in the smaller value
$c_{2.0} = 512 \pm 11$ ppm.
A similar calculation using (4.3) results in $M_{\text{em}} = (898 \pm 43)$ GtC with a
budget $\Delta M_{\text{em}} = (195 \pm 43)$ GtC, comparable to the value 237 GtC shown
in the table below.

The following diagram, taken from Table 8, p.2663 in [20], contains more
current **data from June 2025** for several avoidance probabilities:

**Estimated remaining carbon budgets in GtC
as of the beginning of 2025**

Avoidance probability	17%	33%	50%	67%	83%
1.5 °C	87	54	35	22	8
1.6 °C	169	114	84	65	44
1.7 °C	248	174	134	106	79
2.0 °C	488	357	286	237	188

A higher avoidance probability of limiting warming corresponds to a lower remaining carbon budget.

Here is a graphical representation:

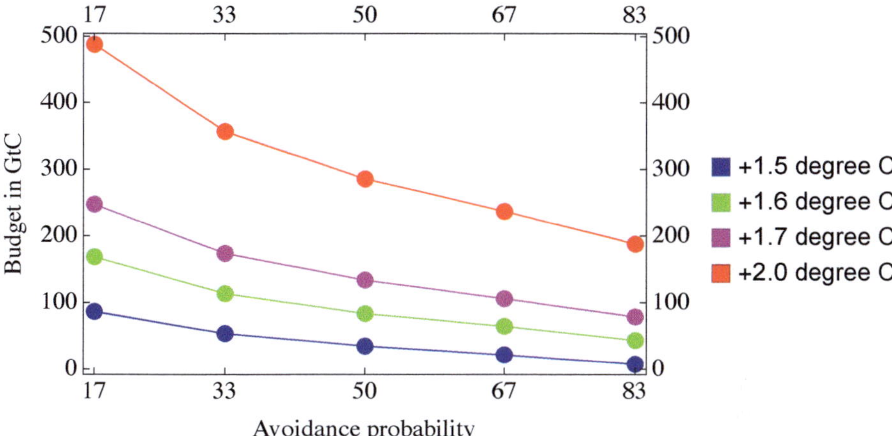

The remaining number of years until the budget is used up can be obtained by dividing by 11.4. It merely corresponds to a different scale of the diagram.

4.4 Decreasing percentage emissions

If the emission quantity, starting from the value of 11.4 GtC for the year 2024, were to decrease by $p\%$ annually, then in the following n years, starting with 2025, the total emissions quantity in GtC would be

$$A_n = 11.4 \cdot (m + m^2 + m^3 + \ldots m^n) \quad \text{where} \quad m = 1 - \frac{p}{100}.$$

This is a *geometric series* (see Appendix).

The *Emission Gap Report of UNEP (UN Environment Program)* from October 2024 concluded: „The global emissions would need an **annual drop of 7.5%** from 2025 through 2035 to get us back on track for 1.5 °C. Cuts of 37% are needed by 2030 ($0.925^6 = 0.63$), 58% are needed by 2035 ($0.925^{11} = 0.42$)."

From the beginning of 2025 until the end of 2035, the accumulated emission would be $A_{11} = 81$ GtC. But the budget of 35 GtC for the avoidance probability 50% is much smaller. And also the budget of 65 GtC in [21] is smaller. **Consequence: The report eight months back was too optimistic for the limit temperature of 1.5 °C!**

Therefore, we refer to the higher limit temperatures of **1.6 °C** and **1.7 °C**.

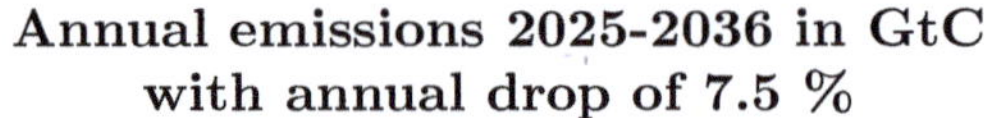

**Annual emissions 2025-2036 in GtC
with annual drop of 7.5 %**

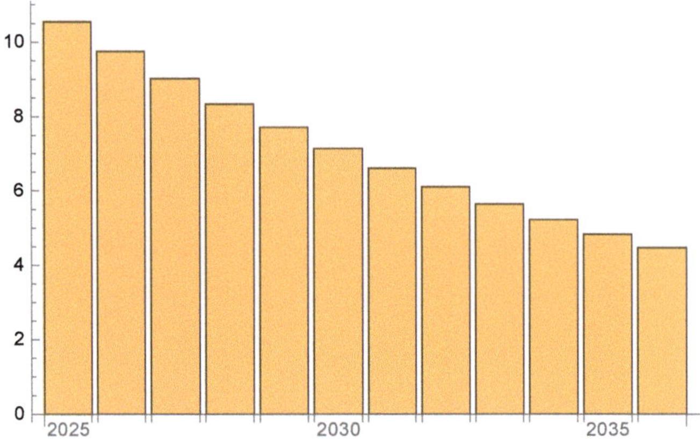

with their

Accumulated emissions in GtC

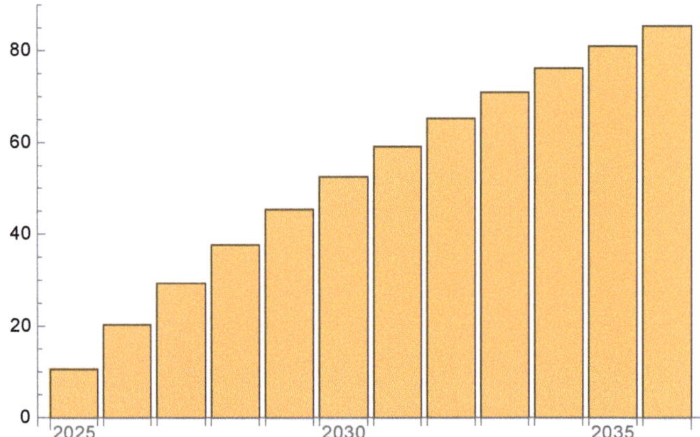

- **Limit temperatures 1.6 °C with 50 % avoidance probability:**
  For the duration of the 12 years from the beginning of 2025 until the end
  of 2036, the total emission would be 85 GtC with an emission in 2035 of
  $11.4 \cdot m^{11} = 4.8$ GtC.
  This means that the budget of 84 GtC in the previous diagram is completely used up. An abrupt stop of the emission of 4.8 GtC had to be
  realized from 2036 on!
  **Implementing an annual percentage decrease of 7.5% and after
  2036 a complete stop is a massive challenge!**

  Reason: The emission during the COVID year 2020 relative to the years
  2018 and 2019 fell only by 6% (see diagram on page 26)!

- **Limit temperature 1.7 °C with 50 % avoidance probability:**
  The budget for this case is 134 GtC (see previous table) or 160 GtC (see [21]).
  If the emission quantity were to decrease by 2% ($m = 0.98$), then the total emissions for the period from the beginning of 2025 until

  - the end of 2037 would be $A_{13} = 129$ GtC. **An abrupt stop of the emission of 9 GtC had to be realized from 2038 on!**
  - the end of 2040 would be $A_{17} = 162$ GtC. **An abrupt stop of the emission of 8 GtC had to be realized from 2041 on!**

4.5 Linearly decreasing emission and greenhouse gases

Let us consider the following text with the subsequent diagram taken from [41]: „In the future, an enormous amount of CO_2 will have to be removed from the air in order to slow down global warming. We are still far from it.

A scenario with an approximately linear decrease is discussed in the diagram (blue line). It shows how the annual pollutant emissions in billions of tons of CO_2 equivalents ($GtCO_2$) per year have to decrease in order for the climate target of 1.5 °C to be met by 2030.“

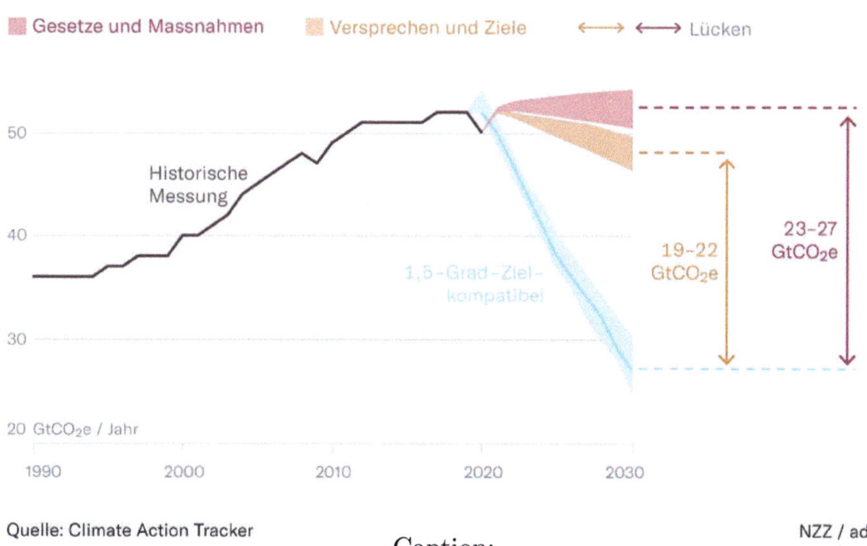

Caption:
pink: laws and measures
yellow: promises and targets
arrows indicate gaps
black curve: historical measurements

The scenario is **incompatible** with the current promises and targets outlined in orange (these are mainly met by purchasing certificates), not to mention the laws and measures marked in pink: the two gaps are way too big!
Despite promises regarding the climate, the target of 1.5 °C is widely missed.

The next diagram from [75] shows that from 2010 onwards the CO_2 share (region marked in blue and orange) accounts for 75% of all greenhouse gases. The amount of CO_2 equivalents was a third higher than the amount of CO_2. Hence for 2021, with the total emissions being 36.8 $GtCO_2$, the amount of CO_2 equivalents total $36.8 \cdot 4/3 = 49.1$ $GtCO_2$. This corresponds to the peak value of the black curve in the preceding diagram.

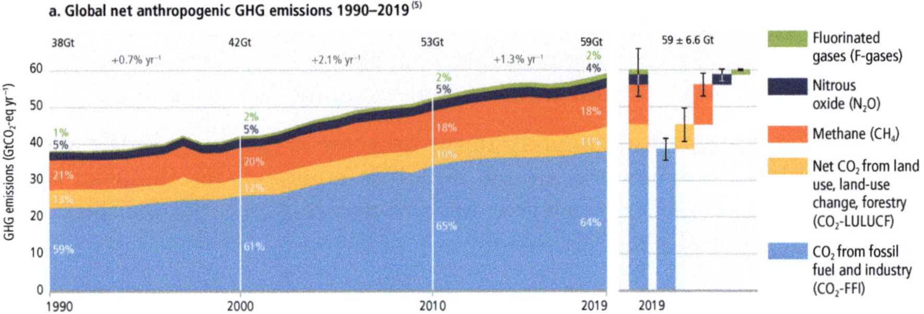

Carbon neutrality or **net zero** is the balance between the amount of greenhouse gases emitted and the amount removed from the atmosphere in greenhouse gas sinks (e.g. oceans and forests).
Man-made efforts such as reforestation and appropriate technologies are indispensable in order to achieve net zero.

The President of the École Polytechnique Fédérale de Lausanne (EPFL), Professor Martin Vetterli, said in an interview with the *Neue Zürcher Zeitung* on June 1, 2023:
„If we can go to the moon or to Mars, we can probably reach net zero! There are areas where decarbonization is difficult, such as in the production of aluminum, steel and cement. But apart from those, low-emission technologies are available for almost all areas of the economy.“

Chapter 5
Causes of global warming

Generally, the main cause of global warming is the use of large amounts of fossil fuels (coal, natural gas, oil and peat) for mankind's energy needs.

From [55]: In 2021, 81% of global CO_2 emissions were caused by the G20 countries (among those are China, USA and India) with 62% of the world population. The largest emitters relative to the total global emissions were

- China with a population of 1.4 billion people at 31%,

- USA with a population of 330 million at 14%,

- India with a population of 1.4 billion people at 7%.

5.1 The various carbon polluters

1. The **CO_2e pollutant emissions per capita** in 2021 for India (I), Sweden (S), China (C), Germany (G), USA (U), Qatar (Q) roughly obey the following proportion:[1]

$$Q : U : G : C : S : I = 14 : 6 : 4 : 3 : 2 : 1.$$

2. The **energy consumptions per capita** in 2021 obey approximately the following:

$$Q : U : G : C : S : I = 44 : 14 : 7 : 5 : 11 : 1.$$

   G and S swap places in the two proportions: The CO_2 pollution of G is double that of S, but the energy consumption of G is only 2/3 that of S. Germany's production using fossil fuels on the one hand and the heating needs in the far north, on the other hand, reflect this.

---

[1] From: *The climate debate suffers from denial of reality,* by Toni Stadler in the *Neue Zürcher Zeitung* of November 12, 2021.

© The Author(s), under exclusive license to Springer-Verlag GmbH, DE, part of Springer Nature 2026

A. Fässler, *Mankind's Problem: Climate Change,*
https://doi.org/10.1007/978-3-662-71846-9_5

3. **Global CO$_2$ emissions by sectors.**
   The following proportion holds:

   Energy production:Agriculture:Industrial process:Waste $= 18 : 3 : 2.5 : 1$.

   The remaining sectors are negligible, accounting for 0.5% of total emissions.

4. **Global emissions by various greenhouse gases.**
   The following proportion holds:

   Carbon dioxide CO$_2$: Methane CH$_4$: Nitrous oxide N$_2$O $= 23 : 5 : 2$.

   The other greenhouse gases make up 2%.

5. According to [55], the global **cement production** in 2022 was about
   4.1 GtCO$_2$. Hence it is responsible for 8% of the global CO$_2$ emissions. In
   1995 it was just 1.2 GtCO$_2$! The reason is the global construction boom
   due to a growing world population and an increasing urbanization.
   Producing 1 ton of cement requires about 110 kWh of electrical energy,
   as much as a 3-person household consumes in 2 weeks.
   There are endeavors to lower CO$_2$ emissions by way of the so-called CO$_2$
   capture technology.

6. The global share of the climate impact of **air traffic** consists of direct
   CO$_2$ emissions alongside nitrogen oxides and water vapor in the higher
   layers of the atmosphere. The IPCC currently estimates the total global
   climate impact at 4.9%. Other sources give values between 3% and 7%.

7. **Ocean shipping** contributes about 2.9% of all global greenhouse gas
   emissions, according to the study of the International Maritime Organization (IMO) in 2020.

8. **Aluminum production** is one of the most energy-intensive industrial
   processes. The production of 1 ton of aluminum requires approximately
   15 MWh $= 15,000$ kWh of electrical energy, as much as a two-person
   household consumes in 5 years. On average, a car contains 150 kg of
   aluminum.
   Data regarding Switzerland: In 2021, aluminum exports were 73,000 tons
   and imports were 288,000 tons. The production of the import volume
   alone requires 4,300 GWh. This is more than the total electrical energy
   consumed by the Swiss Federal Railways in one year with 3,100 GWh.
   A medium-sized nuclear power plant delivers a net output of approximately 1 GW $= 10^9$ W and would supply energy to produce the imported
   quantity for half a year (24 h/day).

9. In the first 7 months, the **war in Ukraine** emitted 100 million tons of
   CO$_2$, extrapolated over a year, that was 171 million tons of CO$_2$. This
   amounts to about 0.5% of the annual emissions of 37.1 GtCO$_2$.

5.2 World population, energy, water and food securities

In November 2022 the world population exceeded 8 billion. By 2050 it could be close to 10 billion = 10,000 million. Eight countries are responsible for half of the population growth: five African and three Asian.

Africa's population reached 1.5 billion in 2024, having grown more than five-fold since 1960 (with 283 million).

In 2050, according to UN forecasts, around 40% of all newborn babies will be African, and Nigeria will become the third most populous country in the world with 450 million inhabitants, as many as are currently in the entire EU. According to the Nigerian Bureau of Statistics, 2/3 of the population currently live in poverty.

The world population has tripled over the past 70 years.

The fertility rate (birth rate) is a measure used in demographics to indicate the average number of births to a woman over the course of her life.

Some fertility rates for 2023 (Source: Statista):

- globally 2.25 children
- Africa 4.07 children
- Asia 1.88 children
- India 1.98 children
- North America 1.6 children
- Latin America & the Carribbean 1.8 children
- Europe 1.4 children

A rate lower than 2.0 means that in the long run the population is shrinking. However, how much the world population will continue to rise for the time being depends on its age structure and the increasing life expectancy.

The United Nations World Population Prospects (WPP) forecasts a rate of just 1.01 for China in 2024. Although a falling rate is also forecast for India (1.91 for 2030 and 1.8 for 2040), the population will continue to rise for some longer period of time. The reason is that with an average age of 27.9 years, India has the youngest population among all industrial and emerging countries. More information can be found in *Population.UN.org*.

Forecasts for India according to [53]:

year	population of India
2023	1,427 million
2030	1,514 million
2040	1,612 million
2050	1,670 million

In 1960 the Indian population was 451 million.

South Korea: Its fertility rate, the lowest in the world, rose from 0.72 in 2023 to 0.75 in 2024.

The two following graphs represent estimates of the evolving population of the world and in different continents according to the United Nations.

Population in million (logarithmic scale)

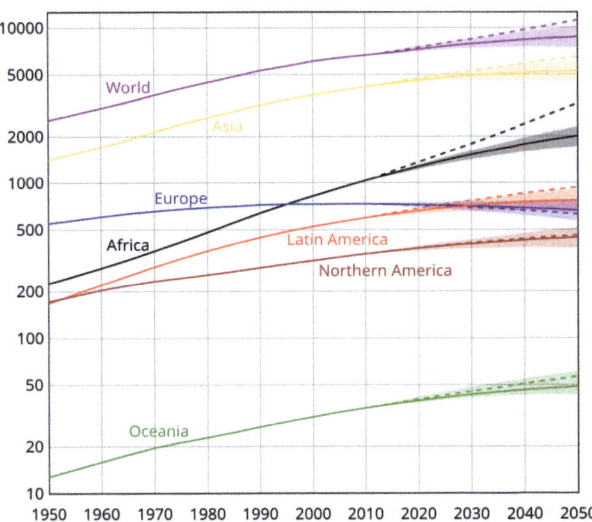

Populations for 1950 (yellow) and 2050 (violet)

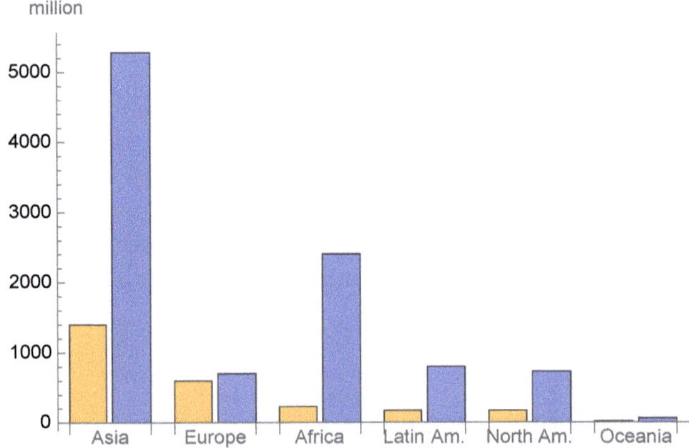

The United Nations projects **the world population in 2050 to be between 9,400 and 10,000 million (10 billion)** with a 95 % confidence interval and a peak in the mid-2080s, reaching around 10,300 million.
Asia remains dominant with a projected 55 % of the world population in 2050, but the population of Africa will increase with a factor of about 10 during the 100 years from 1950 until 2050.

The medical journal *The Lancet* made the following statements under the heading *World population will drastically decrease if birth rates continue to fall*: "Experts expect that the number of people on the planet will reach a maximum of around 9.7 billion in 2064, and will then fall to 8.8 billion by the end of the century."

From [51], [52] and [54]:

Population densities in number of inhabitants per km^2

Continent	in 2020	forecast for 2100
Asia	149	149
Africa	47	133
Europe	34	27
South America & Caribbean	32	32

country	in 2022	forecast for 2050
India	473	562
China	149	137
USA	37	37

Among the most sparsely populated areas are Australia and Canada with less than 4 people/km^2 and Russia with 9 people/km^2.

On the subject of **aging**: Japan, with a population of 123 million at the beginning of 2025, counted 10.4% of people who were 80 years or older and 0.8 per mille who were 100 years or older[2].
About 8 million houses are vacant and numerous villages are abandoned.
Predictions for the year 2050 are that 17% will be 80 years or older and 0.7% will be 100 years or older. Despite aging, the total population relative to 2025 will decrease by 15% to 104 million.

Both the growth of the world's population and a higher standard of living with growing consumption of food and goods will massively increase energy demand in the future.

The goods also include the harmful **production of waste.**[3]
There are about 5 plastic islands in the oceans. One of them, the Great Pacific Garbage Pad (GPGP) between Hawaii and California, covers an area of 1.6 million km^2 corresponding to 4.5 times the area of Germany.

---

[2] Source: *Japan Times* of September 17, 2024

[3] The Czech writer Ivan Klima caused a sensation with his 1991 book [30] *Love and Garbage*, with the gloomy prognosis: "Because nothing disappears from the planet's surface, the consequences of our actions will one day bury us."

The ocean conservation organization *Oceana* estimates that every hour about 675 tons of waste are thrown directly into the sea worldwide. Half of which is plastic. This is a gigantic ecological problem!

According to the Global E-Waste Monitor report of the UN Research Institute and the International Telecommunication Union, there was 82% more electronic waste worldwide in 2022 than in 2010.

So far, more than 500 million people in Africa have no electricity. To cover the growing energy demand in the years to come, more than 20 new coal-fired power plants with an output of about 47 GW could go into operation there. Similarly in Asia!

However, it is urgent to shut down coal and gas-fired power plants and to replace them by expanding fossil-free energy production. Mankind has no other choice if the planet is to be preserved for future generations!

Fossil fuels dominated global energy production in 2023 with a share of 81%.

A promising, robust technology for **heating systems** is the *geothermal energy*, which is available in huge quantities beneath the earth's surface. *Solar systems* and *heat pumps* offer further solutions to avoid fossil fuels.

More and more households are switching to *wood* as a sustainable fuel in the form of logs, wood chips and pellets. However, it is also known that burning wood is associated with considerable *emissions of particulate matter* – even with modern small combustion plants that comply with the latest legal requirements.[4]

The demand for **air conditioning** to cool rooms will increase significantly in hot countries. "It is as legitimate as the claim for heating in cold regions."[5]

The **food problem must be rethought:** Higher meat production and overfishing put a strain on nature and require more energy.

From the *2023 IPCC Working Group II Report*:
"Agricultural productivity in Africa has fallen by 34% since 1961, caused by extreme droughts.
The drastic melting of glaciers is having a major impact on the water balance in the Himalayan region: the Ganges and, above all, the Indus are fed by meltwater. As a result, disappearing glaciers cause water shortage in agriculture and electricity production.

---

[4] This is illustrated by a series of field studies by the Max Planck Institute for Chemistry in 2021: Simple systems, such as private wood-burning stoves, lead to a measurable increase in the concentration of fine dust in their environment and thus endanger health. In the case of larger municipal plants which, in addition to controlled complete combustion, also have an efficient exhaust gas cleaning system and a high chimney, no detectable influence on the local air quality could be measured.

[5] Statement by Toni Stadler, a colonial historian, development expert and publicist, with 25 years in international service in Asia and Africa, to the International Red Cross, the UN and as a Swiss delegate to the OECD.

Various fish species will go extinct due to acidification and depletion of oxygen in the oceans. By 2100, fishing in Africa is projected to fall by about 40%, even if maintaining a temperature increase of 1.6 °C."

Another serious problem is the **earth's water balance**. Typical consequences of climate change on the earth are warming and acidification of the world's oceans (with consequences for the marine ecosystems), the increase in sea levels, the retreat of glaciers, polar ice sheets, Arctic sea ice and snow cover in the Northern Hemisphere, and the thawing of permafrost.

Even a few extreme weather events may cause major social and economic damages. According to the *2022 IPCC Sixth Assessment Report (AR6)*, they are becoming more frequent, more serious and affecting more and more regions of the world.
Lack of or less frequent rainfall, increased evaporation, decreased soil moisture, and increased runoff cause more frequent and severe droughts.

According to the United Nations *World Water Report*, by 2030 almost half of the world's population will be living in regions that have little access to drinking water. Water, being the basis of life, harbors the risk of conflicts, which are exacerbated by the growth in world population. The risk of food crises also increases.

5.3 Mobility

Facts about automobiles:

1. J. Stackmann[6] made the following statement about **cars** in the *NZZ* of September 24, 2022:
   "Public passenger traffic must be expanded. No question. At the same time, the use of cars should change.
   Currently, cars stand mainly an average of 23 hours a day. In addition, they are usually used by one person only. Producing millions of cars so that most of the time they just stand around doesn't fit in at all with a world view of resource sustainability.

---

[6] Director of the Future Mobility Lab of the Institute for Mobility at the University of St. Gallen, Switzerland. Managing Director for Sales and Marketing of various car companies, most recently at Volkswagen until 2020.

The car only makes sense when its benefits are shared. We see the golden way for car retailers in autonomous minibuses for six to eight people.
If there are such offers from companies, private individuals no longer have to buy a car."

2. According to *Statista*, a total of about 1.4 billion cars were produced worldwide in the 16 years from 2008 to 2023. Assuming that the average global lifespan of a vehicle is around 16 years (in Germany around 12 years), there were 1.4 billion cars in use globally as of 2024. The global fleet of all motor vehicles (cars, trucks, buses, motorcycles) rises to about 1.5–1.6 billion. Beijing hosted 6 million cars.

3. If the global vehicle density were to amount to around 500 vehicles per 1000 people in the longer term (USA: 800 cars per 1000 people), the planet would one day be burdened with an avalanche of 4 billion vehicles. According to [31], 2 billion are expected by 2030. It is to be hoped that such scenarios will not occur.
   The situation in big cities has been bad for a long time. Some examples of average traffic speeds: Dhaka: 5 km/h; London: 18 km/h; Jakarta during rush hours: 5 km/h. In addition, the level of air pollution has long been toxic. Impressive pictures of traffic breakdowns in Dhaka, Karachi, Lagos and Kathmandu can be found in [31] on pages 346 and 347.

4. **Today's automobile system, especially in cities, is reduced to absurdity, because of its far too large number.**[7]
   In addition, private transport claims many victims. A total of 2,770 people died in road traffic accidents in Germany in 2023. In comparison, the terrorist attack in New York claimed 2,977 victims.
   Furthermore, traffic noise impairs the quality of life and health of millions of people on earth.

5. **Electric car vs. gasoline car**: In this comparison, all information refers to average values. The *Joanneum Research company in Graz, Austria*, conducted a Life Cycle Analysis (LCA) of a car. Assuming an annual driving distance of 15,000 km during its lifespan of 16 years, the study concluded that the energy-intensive battery production for electric cars amortizes within 3-4 years compared to the gasoline engine.
   This analysis is consistent with the ADAC study, which states that, with a total driving distance of 45,000 – 60,000 km, the two car types are tied.

   In the LCA, the battery production of an electric car accounts for about 115 g CO_2e per kilometer driven if the total driving distance is approximately 240,000 km.
   For the gasoline engine it is 244 g of CO_2e per kilometer driven.

---

[7] Federal Statistical Office of Germany: In 2023, an estimated 4.6 billion of the total of 8 billion people lived in cities, which corresponds to 57% of the world population, with an ascending trend.

The efficiency, i.e. the proportion of energy used for locomotion from well-to-wheel [8] is as follows:

- electric car 65%,
- gasoline car 22%,
- car with fuel cell 20%,
- car with e-fuel 13% (synthetic fuels).

An electric vehicle consumes around 16 kWh. At a price of 0.4 euros/kWh, the energy cost amounts to around 6.4 euros per 100 km travel distance. For a combustion car with a consumption of about 7.4 liters per 100 km and a price per liter of 2 euros, the energy cost is 14.8 euros per 100 km of travel.
The electric vehicle therefore clearly outperforms the gasoline car both ecologically and economically. Other institutes come to similar conclusions with their analyses.

6. From raw material extraction to final assembly, an average of about 400,000 liters of water, i.e. 400 tons of water, are used for manufacturing a car.

Comparison of different means of transport:

1. Average pollution rates per person and kilometer in g CO_2e are as follows:[9]
   - flights: domestic 255, long distance 150,
   - car with one person 180,
   - bus 105,
   - train 40.

2. In freight transport, trucks exhibit an average pollution rate per ton that is six times higher than that of trains.

3. Motorways and highways need much more cultivated land than railways.

4. The *MyClimate* foundation, based in Zurich, calculates a pollution rate of 3 tons of CO_2e for a return flight Zurich-Shanghai, which is equivalent to the annual pollution rate of a car driving 12,000 km.

5. Cruise ships cause high carbon dioxide emissions. A 7-day Mediterranean cruise generates around 1.9 tons of CO_2e per person.

---

[8] The entire chain of effects for the locomotion from the extraction of energy (well) to its provision and its conversion into kinetic energy at the wheels is taken into account.
   Source: https://www.google.ch/search?q=wirkungsgrade+auto

[9] From *Statista:* carbon-footprint-of-travel-per-kilometer.

Possibilities for ocean shipping to reduce CO_2 emissions are

- speed reduction instead of waiting for days at the destination port for cargo to be unloaded,
- optimizing traffic routes,
- eco-friendly fuels.

5.4 Cutting down forests

The tropical rain forests are of crucial importance for the climate.
In its *2020 State of the World's Forests* report, the *Food and Agriculture Organization of the United Nations FAO* estimates that there are approximately 18.3 million km^2 of tropical forests worldwide. This amounts to about 12% of the total land area on earth, covering 149 million km^2.

The three largest contiguous rainforest regions with a total area of approximately 13.4 million km^2 (9% of the land area) are the following:

- The **Amazon rainforest** covers about 8 million km^2, that is, about 5% of the earth's land area (of which 5 million km^2 are in Brazil).
  It has an overwhelming variety of species:
  Around 40,000 plants; 427 mammals; 1,300 birds and 3,000 fish species.
  Since 1990, more than 400,000 km^2 of forest (roughly the area of France) has been cleared.
  The main drivers of deforestation are: cattle farming, soybean cultivation, harvesting tropical timber, palm oil plantations, timber plantations for paper production, infrastructure projects (for instance, dams and mining of raw materials such as oil or gold).
- The **rainforest in the Congo Basin**, with an area of 3 million km^2, absorbs more CO_2 than Africa emits. It hosts the world's largest tropical peatland.
  Its biodiversity is impressive: 10,000 plant; 400 mammal; 1,000 bird and 1,300 fish (in the Congo River system) species. The Democratic Republic of Congo is auctioning rainforest, and thus threatening the global climate. We are talking about millions of hectares for oil and gas production!
- **Rainforests in Southeast Asia** take an area of about 2.4 million km^2. Since the turn of the millennium, Indonesia has lost 27 million hectares. This means 270,000 km^2, equivalent to about three quarters of the area of Germany, of forest area are lost to palm oil plantations.

The valuable peat soils, on which the rainforests have grown over many centuries, are drying up due to clearing.

5.5 Methane emissions and carbon storage

According to the International Energy Agency, methane is currently responsible for around 30% of the global rise in temperature. One kilogram of methane contributes 26 times as much to the greenhouse effect as one kilogram of CO_2. Human activities are responsible for about 60% of methane emissions, with the remaining 40% coming from natural sources such as wetlands.

Methane emissions in 2017 (from [67]):

- "Around 8 $GtCO_2e$ in the form of methane were emitted worldwide, representing 17% of all greenhouse gases emitted.
- About 43% of methane were emitted in China (18.7%), USA (8.1%), India (6.3%), Brazil (5.1%) and Russia (4.7%).
- Around 45% of all methane emissions come from agriculture.
- Energy use makes up 37% and waste management 18% of all methane emissions.

In 1957, global methane emissions were half of what they were in 2017."

A statement from the Neue Zürcher Zeitung:[10]
"Cattle are a climate policy challenge. The methane and nitrous oxide emissions account for 80% of all emissions in agriculture. In the EU they make up 10% of all greenhouse gases. Measures in the EU to reduce methane emissions are needed."

Methane makes up more than 90% of the gas in the North Stream pipelines. From [1]: During the four leaks in the Baltic Sea gas pipelines North Stream 1 and North Stream 2, the estimated escaped amounts of methane were between 56,000 and 155,000 tons. Provided with the factor 26 (see beginning of this section), the values are between 1.5 and 4 million tons CO_2e.

For comparison: The oil and gas industry loses annually 82 million tons of methane through fugitive emissions, incomplete flaring and venting, i.e. around 220,000 tons per day.

The amount of carbon stored in the soil is huge: about three times that absorbed by forests and plants. Even if only a small percentage of carbon is emitted into the atmosphere as a result of the warming of the soil, this further pollutes the earth's surface and atmosphere.

Stored carbon in the soil plays a crucial role in the earth's natural carbon cycle between atmosphere, land and ocean, and it is sensitive to increases in the atmospheric CO_2 concentration.

A higher soil temperature enables living microbes to accelerate the microbial decomposition of organic material. According to [58] methane emissions from wetlands have roughly doubled over the past 20 years.

---

[10] From Planet A in the *NZZ* of December 8, 2022.

Chapter 6
Effects on glaciers

Preliminary: Some glacier physics

Sufficiently recurring snowfalls and cold are necessary prerequisites for the formation and existence of glaciers. The fallen snow thickens multiple times to firn. The firn layers increase the pressure over time and sometimes cause liquefaction and refreezing.

Finally, glacial ice with a density of about 0.91 g/cm^3 is formed. Due to small air pockets, this value is slightly smaller than that of compact ice with a density of 0.917 g/cm^3.

Although ice is incompressible, it behaves like a viscous mass under high pressure caused by its own weight.

The flow rate of the Morteratsch Glacier in Switzerland with an ice volume of 0.7 km^3 in 2022 was around 120 m/year.

The Jakobshavn Glacier in Greenland, 250 km north of the Arctic Circle, has the highest annual flow rate at 7 km/year. It is about 4 km wide and 80 km long.

The following picture was taken from a small icebreaker for about 50 tourists. It was not possible to get any closer to the glacier tongue because of the compact layer of ice.

Jakobshavn Glacier, distance about 2 km, photo by A. Fässler 2015

A. Fässler, *Mankind's Problem: Climate Change*,
https://doi.org/10.1007/978-3-662-71846-9_6

The Ilulissat Icefjord lying in front of it is strewn with icebergs of various sizes, formed by the glacier's calving. The large ones are as tall as skyscrapers and sit firmly on the ground. Note that only about 10% of an iceberg is visible.

Every year more than 35 km^3 of ice fall into the fjord with chunks as gigantic as 1.5 km^3, equivalent to the volume of 1,500 single-family homes!

An important aspect: Glaciers are huge freshwater reservoirs.

Iceberg with moraine inclusion in Ilulissat Bay, photo by A. Fässler 2015

6.1 Arctic

Arctic sea ice: As frozen seawater, it is saline. In winter it covers a large part of the Arctic Ocean around the North Pole and northern areas of Europe, Asia and North America, mostly in the form of floating ice sheets.
From [61]: Its areal extent fluctuated greatly. In 2022 it amounted to 14.7 million km^2 on March 1st, and 4.7 million km^2 on September 15th.

Melting Ice Information:

- The average annual ice volume in 1984 totaled 20,000 km^3. And within 34 years, by 2018, it decreased by half to 10,000 km^3.
- Between the Svalbard Mountains and the North Pole, the mean ice thickness decreased from 2.5 m in 1991 to 1.95 m in 2001, i.e. by 22%.

- The geographic North Pole is occasionally ice-free in summer. The answer to when the entire Arctic Ocean becomes almost ice-free in the summer for several years in a row, depends on various boundary conditions. It is impossible to predict precisely, but from the middle of this century on, is a very plausible estimate.[1]

Image of arctic sea ice:

Photo from 2012: NSICDC/ Julienne Stroeve
Not only the area, but also the thickness of the ice floes diminishes.

The melting of the Arctic sea ice is not only a consequence, but also a critical factor of the climate crisis:
Due to the snow cover, the ice reflects up to 90% of the solar radiation and cools the Arctic region through its high albedo.
The sea water, released by the melting ice, reflects only 10% of the solar radiation. As a result, the Arctic waters warm up more quickly and the ice melts even faster. A threatening positive feedback is the consequence, influencing the weather-determining jet stream and thus endangering climate stability.

The **water temperatures of the Arctic North Atlantic** were higher than at any other time in the past 2,000 years. On 31 August 2023, the temperature was 25.2 °C. More information can be found in [62] and [11].
The average **increase in the temperature of the atmosphere** was 4 °C relative to the pre-industrial age. Thus it was higher than in the rest of the

---

[1] Statements by physicist Lars Kaleschke from the Alfred Wegener Institute for Polar and Marine Research.

world. For comparison, let us again consider the data of 2022: According to the IPCC, the temperature increase averaged 1.24 °C globally and 1.54 °C over the land.

Iceland ice: In 2022 the volume was 3,500 km^3 and the surface was 11,000 km^2. The largest glacier has an area of 8,100 km^2. The Icelandic island has an area of 103,000 km^2.

Greenland Ice Sheet: The following data refer to the year 2022: Average thickness 1.5 km, maximum thickness 3.4 km, area 1.7 million km^2, volume 2.6 million km^3 (9% of the total global ice quantity).

From *Wiki.bildungsserver.de*:
The average mass loss over the period

- 1992–2018 was 148 Gt/year (the record year being 2012 with 464 Gt),

- 2003–2016 was 255 Gt/year.

With 532 Gt in 2019 the loss significantly exceeded that of the record year!

Facts:

1. According to the Danish weather authority DMI, temperatures in the north of Greenland exceeded 20 °C in the summer of 2021.
2. According to [62], rain was measured for the first time in August 2021 at an elevation of 3210 m with a temperature of 0.5 °C. It was 15 °C above the long-term average in August 2021!
3. The earth's surface area is 510 million km^2. Of this, 71% is covered with water. Thus the sea surface claims globally 362 million km^2. The complete melting of the Greenland ice sheet would result in a sea level rise^2 of $0.91 \cdot 2.6$ million km^3/362 million km^2 ≈ 6.5 meters.

6.2 Antarctica

The **ice sheet** around the South Pole sitting on the land and ocean floor covers an area of 11.9 million km^2. It has roughly the same area as Europe. Its maximum thickness is 4.9 km and its average thickness is 2.1 km.

In addition, the floating ice shelf has an area of 1.6 million km^2 and a thickness of up to 1 km.

The gigantic volume of the ice sheet totaled 26.5 million km^3 in 2022, which is about 10 times the size of the Greenland ice sheet and nearly 90% of the total amount of global ice.

---

^{2} 1 Gt $= 10^9$ tons corresponds to about 1.10 km^3 due to the ice density of 0.910g/cm^3.

The 2021 *Sixth Assessment Report of the IPCC* states that the current Antarctic ice sheet lost around 2,670 Gt of ice from 1992 until 2020, i.e. 2,940 km^3. It has thus contributed 8 mm to the global sea level rise.
According to the data from [46] the mean annual ice loss was determined with the aid of satellite observations:

- (40 ± 9) Gt/yr during 1979–1990,
- but already (166 ± 18) Gt/yr during 1999–2017.

Contrary to earlier assumptions, East Antarctica and the high mountains are also melting somewhat. The flat western part is melting rapidly and increasingly so. The melting occurs because of the warming of the sea from below, in contrast to the Alpine glaciers, which melt on their surface.

From [8]: "In 2022, because of the man-made greenhouse effect, the oceans were warmer than they had ever been since measurements began in the late 1950s. The energy stored in the oceans down to a depth of 2 km has increased by 11 ± 8 zettajoules relative to 2021 (1 zettajoule = 10^{21} joules). This is equivalent to about 100 times the amount of energy spent in global electricity generation in 2022."[3]

From [52]: "The so-called *Circumpolar Current*, a huge ocean current around Antarctica, causes additional warming of the sea water. The warming is accelerating and probably penetrating the south from the north. In large parts of Antarctica this is not yet a major problem, because the warm water of the Circumpolar Current is blocked by the continental shelf (that is, the shallow seabed near the coast). But at one point the warm water has found a loophole that poses greater dangers to the entire earth: at Thwaites Glacier[4] in West Antarctica."

From [61]: The summer melt of the Antarctic sea ice in February 2023 reduced its area from 18 million km^2 by 2 million km^2 (about half the size of the EU) and set an all-time record.
Copernicus Climate Change Service: In February 2025, the sea ice extent was about 26% below the 1991–2020 average (and in the Arctic it was about 8%).

6.3 Thwaites Glacier

It is located in West Antarctica. In the time span 2010 – 2025, its area was 192,000 km^2 (about the size of Great Britain or Florida) with an estimated volume of $(483,000 \pm 6)$ km^3 according to *antarcticglaciers.org*. At places its ice is more than 2 km thick (Source: Alfred Wegener Institute AWI 2024).

---

[3] The global electrical energy produced in 2021 was 30,235 TWh. As 1 Tera = 10^{12} and 1 Ws = 1 J (joule), the latter amounts to 30,235 · 10^{12} W · 3,600 s ≈ 1.1 · 10^{20} J.

[4] Named after glaciologist Fredrik T. Thwaites in 1967.

Its bed reaches a sea depth of more than 1 km. Thwaites Glacier holds back the gigantic volume of 26.5 million km^3 of Antarctic Ice Sheet, which has a thickness of up to 4.9 km.
At its front it consists of a floating sheet of ice, the so-called *ice shelf*, which is up to 1 km in thickness.

The following image was taken from Wikipedia:[5]

Tongue of Thwaites Glacier (160 km long, 30 km wide)

In 2002, part of the ice shelf, 85 km long, 105 km wide and 200–400 m thick, broke away (Named Iceberg B-22) into the Amundsen Sea!

The two main reasons for the increased calving and the accelerated retreat of the glacier front towards the mainland are:

I. Large parts of the West Antarctic ice lie below the sea surface. Therefore one speaks of a **maritime ice sheet**. It is washed under by warm water (the temperature is just above the freezing point) and is therefore melting from below, because the Antarctic atmosphere has icy temperatures with an average temperature of around $-55\,^\circ$C.

II. The seabed is **retrogradient**, i.e. it slopes toward the interior of the land (usually it slopes outward). The retreat of the glacier wall is equivalent to the retreat of the so-called base line. It is defined as the area where the maritime ice sheet (also called inland ice or continental ice), which sits on solid ground and is partially submerged, transitions into the free-floating ice shelf. This is the line where glacial ice last touches the ground and from where it begins to float.

---

[5] Source: https://en.wikipedia.org/wiki/Thwaites_Glacier

Since the glacier wall is rising due to the retreat (currently with about 2 km/year) and due to the retrograde inclination, increasing masses of ice fall into the sea as a result. The glacier is calving more quickly and melting away faster. It is a self-reinforcing feedback process.

If the glacier loses contact with the ground on some piece of land and floats, this accelerates its flow rate and hence its melting over the next few decades. In this context one speaks of a **tipping point**. A chain reaction threatens to set off.

The following diagram is taken from [4]:

Recession of the Pine Island Glacier (near Thwaites Glacier)

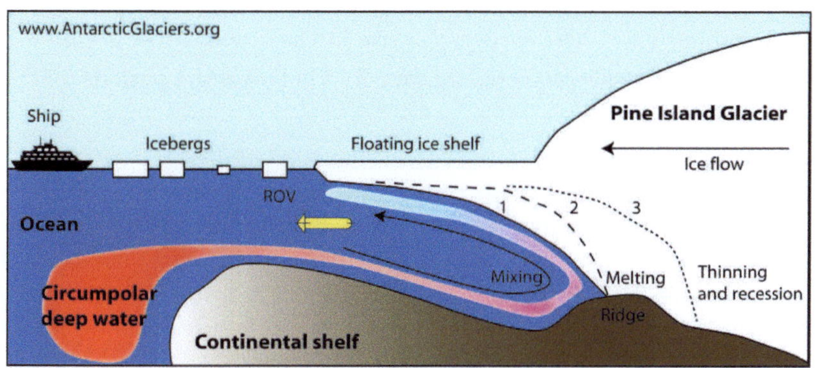

1. Early 1970s. Pine Island Glacier is grounded at a bedrock ridge.
2. Warm, inflowing Circumpolar Deep Water melts the base of the glacier. The glacier steepens and accelerates.
3. Present day, observed by a remotely operated vehicle (ROV). Glacier is thinning and receding.

Warm Circumpolar Deep Water is penetrating beneath the ice shelves of Pine Island Glacier and Thwaites Glacier.

You may consult an informative video under [59].

The Thwaites Glacier acts like a brake pad for the gigantic mainland shield. Its melting would mean that the sea level rises approximately[6]
$$0.9 \cdot 483,000 \text{ km}^3/362 \text{ million km}^2 \approx 1.2 \text{ meters.}$$
But if the Thwaites Glacier breaks away, parts of the gigantic mainland shield will gradually flow into the sea. This would have serious consequences for many megacities near the coast. The OECD warned that coastal flooding could affect 40 to 150 million people by 2070.

If the Antarctic ice sheet with a volume of 26.5 million km^3 were to melt completely, this would result in a sea level rise of no less than
$$0.9 \cdot 26.5 \text{ million km}^3 /362 \text{ million km}^2 \approx 66 \text{ m!}$$

It is not for nothing that the Thwaites Glacier is also referred to as the *Doomsday Glacier*.

---

[6] *antarticglaciers.org* gives a sea level rise of 65 cm.

Ample information can be found in [12]:

1. Between 2006 and 2018, the average sea rise was 3.7 mm/year, during the 20th century it was 1.5 mm/year.

2. Probable sea level rises by 2050 and 2100 for different greenhouse gas emissions in meters:

greenhouse gas emission	sea rise 2050	sea rise 2100
very low (SSP1-1.9)	0.28 to 0.55	0.37 to 0.86
low (SSP1-2.6)	0.32 to 0.62	0.46 to 0.99
medium (SSP2-4.5)	0.44 to 0.76	0.66 to 1.33
high (SSP5-8.5)	0.63 to 1.01	0.98 to 1.88

Throughout the 19th century, the rise in the sea was just 2 cm.

3. The ice sheet loss rate increased four-fold between the periods 1992–1999 and 2010–2019.

The *German Climate Consortium* states that even after net zero, the sea level will continue to rise for several centuries due to the inertia of the climate system. Our actions today will, therefore, have consequences far into the future.

From [74]: Ice melting could have catastrophic consequences for more than a billion people in coastal regions as early as 2050, predicts *the September 2019 UN special report*. People all over the world will have to leave their homes due to rising sea levels.

Statement by glaciologist Ted Scambos (University of Colorado):
"The future of the earth depends on Antarctica, not on Mars."

Numerous documents and pictures can be found on the Internet under the keyword Thwaites Glacier.

In a research experiment carried out in 2019, using an underwater robot called Rán, the following image (from [49]) was taken to measure the seafloor at the tip of the glacier. It shows an extremely regular structure of parallel ridges 7 m apart.

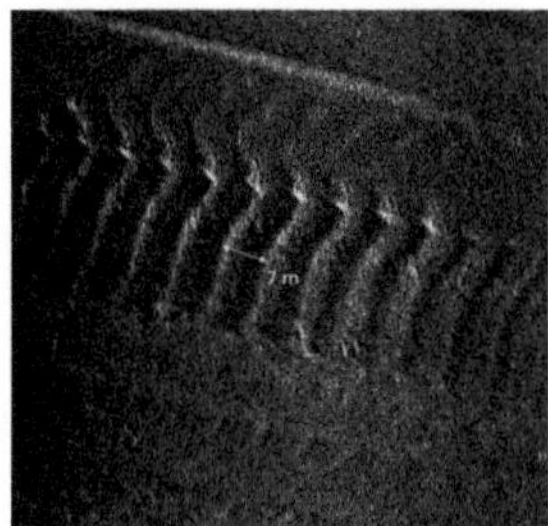

Image credit: Ali Graham

Researchers have discovered more than 160 such ridges. The question arises as to the cause of this extremely regular structure. One explanation put forward is that the tides could have caused this. From my point of view, this is extremely questionable, given the weight of the glacier.

The interdisciplinary theory of **symmetry breaking** (see [22] and [18]), which will not be discussed here, may provide at least a partial explanation.

According to the study [50] led by Britney Schmidt, the main trigger for the collapse of the ice shelf could be the melting of the ice along cracks found in the ice and along the ridges. Warm, salty seawater penetrates through them and widens the cracks and spaces between the ridges.

6.4 All glaciers outside Antarctica and Arctic

In Table 1 of [19], all 215,000 glaciers with a combined volume of 158,000 km^3 were determined also with regard to geographic distribution using a computer model.

The most important percentages of glaciers by volume are (with the number of glaciers in brackets):

- 38.1% (2,750): Antarctica and Subantarctica (excluding Antarctic Ice Sheet),
- 11.3% (19,300): Periphery of Greenland,
- 9.5% (1,070): Russian Arctic,
- 6.6% (12,000): Canadian Arctic (North and South),
- 5.4% (27,100): Alaska,
- 4.1% (1,600): Svalbard,

with a total share of 75%.

On the other hand, the portion of glaciers in Switzerland at the end of 2024 was extremely small, with an estimated total volume of 46 km^3 and an area of 960 km^2. In 2024, the Swiss glaciers experienced a 2.5% loss, equivalent to approximately 1.2 km^3 of ice — comparable to the volume of Lake Biel. In 1999 the total glacier volume in Switzerland was at least 74 km^3 and that of all alpine glaciers was estimated at 80 km^3.

According to the *EU Climate Change Service Copernicus*, in 2022 more than 5 km^3 of glacier ice in the European Alps melted, more than ever before. The volume equals that of a cube of sides 1.7 km, which is five times the height of the Eiffel Tower!

According to [39] the loss of the total volume of V=158,000 km^3 of ice amounted to about 293 km^3/year.

In the January 2023 issue of the journal *Science*, [47], global forecasts until 2100 are made about all glaciers worldwide:

- Even if the climate target with an average warming of $1.5\,°C$ were met, the glaciers would lose $26 \pm 6\%$ of their mass, resulting in a sea rise of 90 ± 26 mm.
  The following computation confirms this:
  Let A = area of sea surface = 362 million km^2 and $V = 158,000$ km^3. Then $0.26 \cdot V/A$ is the mean thickness of the melted ice sheet. Since water is denser than ice, we multiply this by 0.91 in order to get the thickness of the corresponding water layer, i.e. the rise in the sea level.
  Result: 103 mm.
- According to the 2021 UN climate conference *COP26*, where a temperature increase of $2.7\,°C$ was forecast, the sea would rise (115 ± 40) mm in response to the extensive disappearance of glaciers in the central latitudes of the earth.

As an example from the Alps, let us consider the

Morteratsch Glacier in the Engadin valley, Switzerland

It is a so-called valley glacier at an elevation of between 4,000 and 2,000 m above sea level. In 2022 it was just 6 km long with an area of about 15 km^2 and an average width of 2.8 km.

The following two photos by Jürg Alean and Michael Hambrey were taken with thanks from *www.swisseduc.ch/glaciers/*.

Upper photo: In 1985 the glacier front was steep and convex, which is typical.
Lower photo: By 2015, the glacier tongue retreated by more than 700 meters.

The retreat of the glacier tongue within the 10 years from 2000 to 2010 was 315 m, from 1910 to 1920 only 10 m (from [45]).

Chapter 7
Other effects and dangers of global warming

7.1 Extreme weather

They are menacing and endanger human health. Heat, bush and forest fires, storms, drought, heavy rain and floods are increasing in frequency and intensity all over the world.
Extreme weather is becoming more frequent on all continents.

In Australia, devastating bushfires raged in 2019 and 2020: 3 billion animals, including 143 million mammals, died or were driven from their habitat.

In Germany in July 2021 there was an unprecedented flood disaster in the Ahr Valley in Rhineland-Palatinate. 135 people died, hundreds were injured and large parts of the valley were devastated. Cost: 35 billion euros.

According to the *EU Climate Change Service Copernicus*, the earth's global surface temperature—compared to pre-industrial times—was 1.68 °C higher during the period from June 2023 to March 2024. The latter 10 months in a row have all been warmer than ever before since measurements began.
The warmest month to date worldwide was May 2024 with a global mean temperature of 15.91 °C. This was 0.65 °C above the 1991–2020 average and 0.19 °C above the previous high of May 2020.

In 2022, more than 1/3 of the European continent suffered from drought and temperatures in Europe rose around twice as much as the global average.

But the world's oceans are also warmer than ever, according to the data of the *American National Oceanic and Atmospheric Administration NOAA*. Since early 2023, the temperature on the surface of the North Atlantic has always been the highest among those measured in the respective seasons. In March 2024 the water was 1 °C warmer than the average over the period from 1991 to 2020 and 0.4 °C warmer than the previous high in February 2024. The highest global ocean temperature ever recorded was 21.1 °C in February 2024. It was 0.7 ° C above the average over the period 1991–2020.

A. Fässler, *Mankind's Problem: Climate Change*,
https://doi.org/10.1007/978-3-662-71846-9_7

Global warming can be seen particularly well in the temperatures of the oceans, because 90% of the additional heat is stored in the oceans. But only part of the high ocean temperatures is explained by the El Niño phenomenon: It results in temperatures rising mainly in the tropical Pacific. But El Niño cannot explain the high temperatures in the North Atlantic. Convincing explanations have yet to be found.

Extreme events in 2022:

1. Flood disaster in Pakistan: At some time, a third of the country was under water, causing 1700 deaths. Estimated total damage: 30 billion US dollars.
2. Australia's east coast experienced its worst flooding since humans can think. Tens of thousands of people had to be evacuated. The city of Sydney was also affected.
3. China experienced its longest and worst heat wave since weather records began in 1961, with record temperatures exceeding $40\,°C$ for weeks across much of the country. Water levels in sections of the Yangtze River reached their lowest levels since records began in 1865. Energy production with hydropower fell massively, but coal-fired power plants were running at full speed. Large harvest areas were destroyed or damaged.
4. Heat waves in Europe including Great Britain: An area of $7{,}500\ \mathrm{km}^2$ was burned, three times the average of the years 2006–2021.
5. Japan: Extreme cold and snowstorms.
6. Africa: While parts of the continent had been waiting for water for years, hundreds of people died after massive floods. Millions of people were affected by the floods.
7. Catastrophic drought in the Sahel from Senegal in the West to Djibouti in East Africa. 346 million people did not have enough to eat. That was about a quarter of Africa's population.
8. Dubai, Oman, India, Iran, Qatar, Pakistan: Heat waves up to $50\,°C$ followed by devastating floods in northern India.
9. Heat wave up to $45\,°C$ and flooding in Argentina and in several states of Brazil.
10. India: The worst heat wave since 1910, dried up fields, fires, garbage dumps burning for days in Delhi.
11. Italy: Up to $48.7\,°C$ were measured in Sicily.
12. $40\,°C$ heat wave in northeastern US including New York and Boston, and a snowstorm of the century in upstate New York with 2 meters of snow.
13. Winter storm in California with 2 meters of snow.

Extreme events in 2023:

1. On the holiday island of Mallorca palm trees were white in March from snow falling more than 1 meter deep. In addition, there were hurricanes and heavy rains.

2. On March 11 already, temperatures up to 31 °C were recorded in Spain.

3. Hundreds of wildfires raged in Canada in June. Large parts of the eastern United States were enveloped in noxious smoke. Warnings were issued in major cities such as New York and Toronto. Residents were advised to stay at home and close their windows. The smoke even reached Western Europe.

4. In June, China issued the highest heat warning level for several regions. 43 °C were measured in Shandong province.

5. Heat waves in the Mediterranean region in July with persistent temperatures above 40 °C and hundreds of fires around the Mediterranean in Italy, Greece, Croatia, Turkey, Spain, Portugal, Algeria and Tunisia. More than 400 fires on the island of Rhodes forced 30,000 people to be evacuated or to flee the fire on foot, including many tourists. The highest temperature ever recorded in Sardinia was 48.6 °C.

6. In July, the median temperature of the water surface of the Mediterranean was 28.71 °C and that of the North Atlantic was 24.9 °C: the highest values ever measured.

7. According to NASA and the EU Climate Change Service, July 2023 was the hottest month on record globally since measurements began in 1880, with temperatures 0.24 degrees above the previous high.

8. In northwest China's Xinjiang Province, a new national temperature record of 52.2 °C was set in Sanbao Village on July 16. The old Chinese record from 2017 was 50.6 °C.

9. In August, the worst fires raged on the island of Maui in Hawaii, USA. More than 100 people died and over 2000 houses were destroyed.

Extreme events in 2024:

1. In April, the Chinese province of Guangdong experienced a flood of the century.
2. In May 2024, the temperature in the small town of Phalodi in the Indian state of Rajasthan reached 50.8 °C.
3. Spain experienced a heat wave in January, i. e. in the middle of winter, in the south and east with temperatures close to 30 °C.
4. On June 6, temperatures in Las Vegas reached 45 °C, a temperature never before recorded so high and so early.
5. In June there was a once-in-a-century flood in southern Germany.
6. On June 13, 44.6 °C was measured in a western Turkish province.

7. In April, maximum temperatures of 44.2 °C were recorded in Thailand's northern Lampang province. They were just below the national record of 44.6 °C the previous year. With perceived temperatures sometimes exceeding 52 °C, there have already been twice as many deaths from heat stroke as in the whole year of 2023.
8. In August the average surface temperature of the Mediterranean Sea exceeded 30 °C for the first time.

Extreme events in 2025:

1. Wildfires in Los Angeles in January resulted in unprecedented damage and economic loss. At least 25 people died. Estimates indicate that the total economic impact could reach between 250 billion USD and 275 billion USD, making it the most expensive wildfire event in U.S. history.
2. Heat waves in Europe occurred in August with temperature peaks of 45 °C in Portugal, Spain and Greece, 42 °C in Bordeaux, 40 °C in Croatia together with intense wildfires.

Droughts and climate change: Climate experts led by ETH's Sonia Seneviratne show the relation between human-caused climate change and the intense droughts of 2022 in the northern hemisphere. According to the study, climate change makes such droughts twenty times more likely.

In **Switzerland**, the seven warmest years since measurements began, were all recorded after 2010.[1] And the winter of 2022/2023 was the one with the least snow for at least 75 years. The average temperature was 2.5 °C above the long-term average. The data from the observation station *Weissfluhjoch Davos 5WJO* at elevation 2536 m are displayed in the following diagram:
The red curve shows the snow depth in cm (with scale markings on the left from 0 to 500 cm at intervals of 100 cm) during the winter of 2022/23 from October 2022 until March 2023.
The black curve shows the average values of the past 32 years.
And the gray area marks the minimum (lower boundary) and maximum (upper boundary) of the average values of the past 32 years.

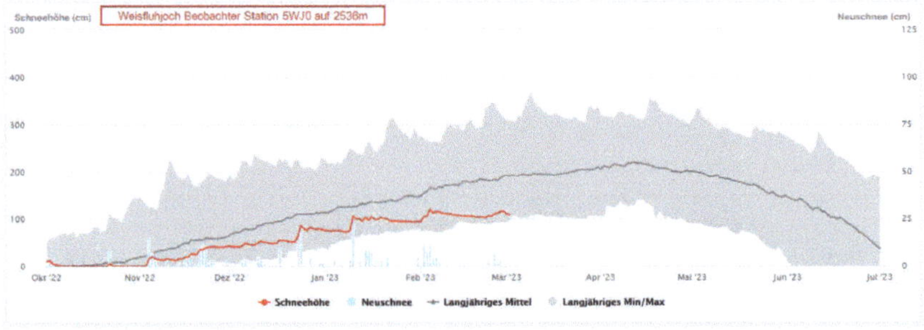

Source: Institute for Snow and Avalanche Research SLF in Davos

---

[1] Neue Zürcher Zeitung NZZ of December 23, 2022.

The values of the red curve are significantly lower than the average values and are sometimes even close to the average minimum curve. For more than 80 years, the SLF has graphically recorded the snow depths measured annually on Weissfluhjoch Davos (in the same format as the previous figure).

Annual maximum snow depths H in cm during 85 years from winter 1937/38 to winter 2022/23

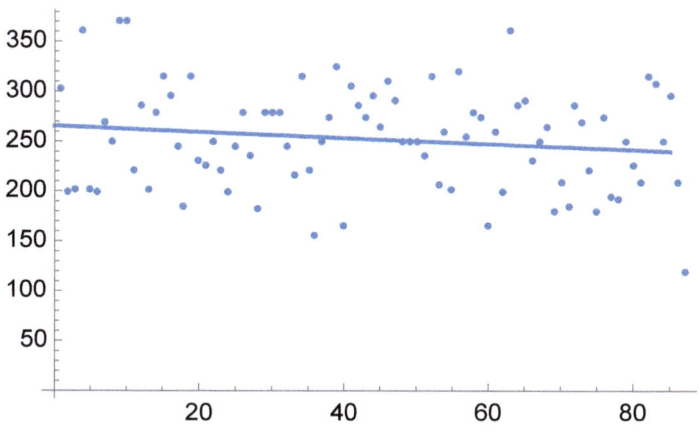

with regression line $H = 266 - 0.30 \cdot \tau$ (τ in years)

The dot on the far left corresponds to the winter of 1937/38. The lowest point on the far right for winter 2022/23 marks the lowest snow depth in all 85 years.

The regression line $\overline{h} = 138 - 0.060 \cdot \tau$ results from the annual mean values $\overline{h}$ estimated by the author.

The winter of 2023/24 was the warmest ever recorded in Switzerland with extremely little snow. Many established ski resorts were forced to stop winter sports operations for lack of snow. Often it rained up to an elevation of over 2000 m above sea level. In many areas even cross-country skiing was no longer possible.

Since 1949, for many years as a child and teenager, the author went skiing in eastern Switzerland at an elevation of 450 m, with plenty of snow in those years. Today it is doubtful whether ski stations at an elevation of 2000 m have a future.

According to the Federal Statistical Office, since 1864 the total amount of precipitation in millimeters of water per year fluctuated practically unchanged around the constant value of 1.2 meters with deviations of up to $\pm$ 30%. The values differ considerably from region to region.

In recent decades, the snow largely disappeared on the Central Plateau, but the average amount of precipitation remained constant. This is another clear indication of global warming.

[39] shows that climate change also increases the risk of snow avalanches.

The **El Niño phenomenon:** It occurs at irregular times every few years and lasts on average 6–9 months.

- Normally, the trade winds blow from the western coast of South America westward towards Australia, as there is a low pressure area over Southeast Asia and a high pressure area over the central Pacific.
  The trade winds cause cool water (part of the Humboldt Current) to be raised from the depths of the ocean off the South American coast. On the one hand, there is drought on the west coast of South America and heavy rain. On the other hand, floods in Southeast Asia are "normal" facts.

- With El Niño, the trade winds weaken and the buoyancy of the cold water decreases and may even come to a standstill. The surface water and thus also the Eastern Pacific off the west coast of South America is warming and the air pressure is falling.
  The increased amounts of evaporated water lead to extreme rainfall and storms along the west coast from South America up north to California. On the other hand, the water temperature off Australia and Indonesia is falling, leading to an increase in pressure. This creates winds in the opposite direction to the trade winds.

From [5] and [6]: As of 2023, the research question remains open as to whether the *frequency* and *strength* of El Niño events are influenced by climate warming. However, it is certain that in the future their *effects* will become stronger, as droughts and floods become more extreme.

7.2 Economic damages

The Potsdam Institute for Climate Impact Research concluded in the journal *Nature*: In the coming 25 years, global warming will reduce the expected global income by 19% compared to a fictional future without global warming. By mid-century, climate change would cause an annual loss of approximately 35 trillion ($35 \cdot 10^{12}$) euros.
The damage in the second half of our century depends on how greenhouse gas pollution develops. If we succeed in reducing emissions, then by around 2050, prosperity would hardly decrease any further.
However, if emissions continue to rise unchecked, greater losses in income must be expected.
The costs of climate protection measures are comparatively low — about 1/6th of the expected losses by 2050.[2]
"Due to the climate's inertia, we have already incurred the damages by 2050, because we haven't acted yet", says Leonie Wenz, an author of the study.

---

[2] *The Neue Zürcher Zeitung NZZ* of April 20, 2024 discusses the topic in more detail under the title *Climate protection may prove to be more cost-effective than global warming.*

That the cost of climate protection is significantly lower than the expected losses, is a strong incentive to reduce emissions quickly. "We must act now to prevent suffering even greater damages in the second half of the century", says the study.

According to the calculations, poorer countries pay a disproportionately high price. Rising temperatures have a particularly drastic impact in Africa and in tropical and subtropical regions.
Only regions in the far north would benefit economically. However, considerably fewer people live there. In essence, the current study in the journal *Nature* agrees with other studies on the economic consequences of global warming.
A few years ago, authors in the journal *Science* came to the conclusion that climate change could cost the US 1.2% per degree temperature increase of the gross domestic product. Another estimate comes to the conclusion that extreme weather events currently already cost 136 billion euros. The authors based their study on historical weather data of the past 40 years as well as on numerical data gathered from studying the economic performance of around 1,600 regions worldwide.

7.3 Biodiversity

The following text is taken from [38]:

"From the ice masses of the polar ice caps to the Great Barrier Reef in Australia and from the coastal forests in East Africa to the rainforest in the Amazon basin, the effects of warming can already be felt everywhere. Studies show that around 1,700 species have begun migrating towards the poles at a recent rate of six kilometers per decade. The climatic zones, i.e. their habitats, are currently shifting more than 7 times faster (50 km per decade). In the future their shifting could be 100 km per decade.

A large number of species are in danger of being left behind. A report by the *Stern* magazine assumes that with a global temperature increase of $2\,°C$ around 25 percent could disappear due to the loss of their habitats. And with that of $3\,°C$ and more, even a third of the known species would go extinct.

Particularly endangered are those species that have no possibility of migrating, such as the coral reefs in the South Pacific, the animals and plants of the polar regions and the Alps. The ability of animals and plants to adapt to climate change will only partially compensate these losses.
A complete extinction of many species can only be prevented by reducing global CO_2 emissions.

According to the Biodiversity Conference in Canada, around 130 species are currently disappearing every day. Just 46% of the world's coral reefs are still healthy.

The importance of coral reefs, which have existed for millions of years, for marine animals and plants is comparable to that of plants and animals for the rainforest!"

Decline and destruction of biodiversity reduces the resilience and productivity of nature. This has disastrous consequences for humanity and the global economy.

7.4 Dangerous climate tipping points

If a tipping point is reached, a state changes abruptly and irreversibly into a new state. Even a small change in the earth system is enough to reach a tipping point.
An example to illustrate this might be the story of Sisyphus, who is assigned to eternally push a large stone up a mountain. But again and again it rolls back. If he were ever to reach the top, this would correspond to the tipping point: a small push would roll the stone down the other side.

Professor Thomas Stocker, climate researcher at the University of Bern and long-time member of the IPCC, commented on the **question of what the dangerous tipping points in the climate** are[3] as follows:
"Understanding tipping points is extremely important for research — and humanity. Their dynamics are not yet fully understood. But there is progress. Since it became clear that CO_2 emissions are rising 100 times faster and 50% higher than in the last 800,000 years,[4] tipping points have become the focus of burning questions."

Tipping points with dangerous global effects could be

- the drying up of the northern branch of the Gulf Stream,
- the melting away of large ice masses in Antarctica,
- the collapse of ecosystems.

But regional tipping points are also increasingly in focus:

- Shifting storm tracks,
- tipping of the monsoon,
- drying out of the Amazon rainforest.

If these systems tip over, resources may become scarce and habitats may be destroyed.

---

[3] Source: *Sonntagszeitung* of July 23, 2023.
[4] The diagram in Section 3.2 applies.

For several years now, researchers have been trying to identify and better understand tipping points in the climate system. Nevertheless, we cannot answer the question. However, critical points in dynamic systems are well-studied objects in mathematics and theoretical physics: Even warning signals are known.

But the climate system is so complex – the lengths and time scales of the relevant processes extend over at least 15 orders of magnitude – that the question will remain a tricky one for the next few years.

Two developments bring us closer to an answer:

1. Thanks to supercomputers, climate models are becoming increasinly fine-grained. This allows a more precise understanding of the processes leading to tipping. The first models with grid width 1 km, covering the entire globe and including the atmosphere and ocean, already exist: dynamic ice flows will also be installed soon. And atmospheric chemistry must also be accounted for. This leads to a better estimate of what can tip over – and when.
   The global research community must now develop the best models to answer this question, essentially a Marshall Plan for climate modeling.

2. The full dynamics of the climate system can only be revealed by a meticulous analysis of climate archives. We reconstruct the greenhouse gases of the past using ice cores from Antarctica.
   The plan is to use this core to advance into an era of faster and weaker ice age cycles – terra incognita. Whether tipping events are also recorded there and what they look like will be a significant key to answering the question.

7.5 Permafrost

Permafrost contains 1,300 Gt to 1,700 Gt of carbon — almost twice as much carbon was 2024 in the Earth's atmosphere.
Permafrost covers 25% of the earth's surface, especially areas in Siberia, Alaska and Canada. They are frozen all year round.

When the permafrost thaws, microorganisms become active and convert carbon compounds stored in the soil into methane, water vapor and carbon dioxide, which in turn increase the greenhouse effect. In Siberia, some railway lines will no longer be traversable because the rails are sinking.
Oil pipelines are becoming unstable and risk leaking. Coastlines erode more and more, houses tumble into the sea.

The thawing of the permafrost has long since arrived, with unfortunate consequences: in 2000, a rock slide hit the small Valais village of Gondo, Switzerland. 13 people lost their lives. Houses were damaged or completely destroyed.

Rock formations in the Alps begin to slide or fall, threatening houses and people. SAC huts of the Swiss Alpine Club have to be abandoned or relocated, as in the case of its 127-year-old hut on the Matterhorn.

On the release of carbon from permafrost:[5]
A total of approximately 100 GtC is found in permanently frozen ground in 2024. By the end of the 21st century, a release of 55–232 GtCe is expected, depending on the IPCC scenario. This will significantly reduce the remaining carbon budget.
If Arctic permafrost were to melt completely, the amount of carbon emissions release would be enormous. Complete thawing with 1,300–1,700 GtC would be catastrophic for climate goals.
Even partial thawing (100-200 GtC) would consume the budget for the limit temperature of $1.7\,°C$ with an avoidance probability of 50%.

7.6 Gulf Stream

The Gulf Stream is part of the ocean circulation in the North Atlantic. It transports warm water on the surface from the Gulf of Mexico towards Europe and causes a mild winter climate there. The surface current with an average width of around 50 km bends eastwards towards Europe. This is due to the predominant westerly winds and the Coriolis force. In the open North Atlantic, the Gulf Stream mixes with cold water from the Labrador Sea and becomes the North Atlantic Current.

The Gulf Stream is the heat pump for Northern Europe. Without it, the climate in Northern Europe would look very different.
There are fears that the Gulf Stream could come to a standstill due to global warming. However, these fears have recently been significantly revised. The following paragraph was taken from the *NZZ newspaper of March 3, 2025:*

„**A collapse of the Gulf Stream system is unlikely.** According to a new study published in the science magazine *Nature*, strong winds in the southern polar region maintain ocean currents in the Atlantic.
The winds were identified as the reason for this, using 34 climate models: The Atlantic circulation is characterized by large masses of water sinking to the depths in the northern part of the ocean because the water is salty and cold, and therefore has a higher density. However, climate change is causing water temperatures to rise and salinity to decrease. Both factors make the water lighter. Scientists therefore assume that the Gulf Stream will weaken in the 21st century.
In the Southern Hemisphere, however, there is an opposing factor: the wind around Antarctica. This pulls water from the depths up to the ocean surface.

---

[5] Private communication from Dr. Jens Strauss, Alfred Wegener Institute AWI, Center for Polar and Marine Research, University of Potsdam.

Climate change cannot harm this wind. It is even assumed that it will become stronger still because the temperature difference between the cold Antarctica and the warmer surroundings is increasing. This powerful pump will keep the Atlantic circulation going in the future, the study says. The simple principle behind it: If water reaches the surface in the southern oceans, water must sink somewhere else. And this happens primarily in the North Atlantic."

7.7 Troposphere and Stratosphere

An increase in greenhouse gases in the atmosphere leads to more heat remaining in the lowest layer, the so-called troposhere, while, according to mathematical models, less heat escapes into the next higher layer, the so-called stratosphere, which extends from altitudes of 15 km to 50 km.
As a consequence, the warming of the troposphere increases, but the stratosphere cools slightly.
In 2023, its cooling was confirmed in [48] by observations and measurements of the global mean temperature using satellites and weather balloons between 25 km and 50 km.

Chapter 8
Philosophy about climate change

Sir Karl Popper (1902–1994)[1] founded the critical rationalism with his work on epistemology, philosophy of science and social philosophy. He formulated the most important criterion for the empirical sciences, namely the **falsifiability: "A scientific hypothesis cannot be strictly proven, but it can be disproved if it is false."**
Only what is formulated in such a way that the hypothesis can be debunked as false is of scientific value. Researchers must strive to refute previous results, i.e. to falsify them. This means that, even if a hypothesis can never be proven or will never be considered ultimately true, one can be all the more sure of it the more stubbornly it defies attempts at falsification.

Popper rejected the inductivist view that a generally valid law can be inferred from observations. Falsificationism therefore assumes that a hypothesis can never be proved, but can be disproved if necessary. This basic idea is older than Popper, it can be found with August Weismann (1834–1914)[2].

Newtonian mechanics provides a relevant example of a successful falsification. For a long time it was considered an indisputable law of nature. It was confirmed by numerous observations and allowed important predictions, for example in astronomy. However, Einstein successfully falsified Newton's principle through physical experiments. He formulated the more comprehensive theory of relativity, which includes classical mechanics as a special case with small velocities (compared to the speed of light).

Quantum mechanics provided another prominent example. It shows that Newtonian mechanics fails in the atomic range. Quantum jumps are incompatible with the continuous processes described by Newton.

The hypothesis of climate change has withstood all attempts at falsification over the past decades!

---

[1] Austrian-British philosopher.

[2] German physician, geneticist, one of the most important evolutionary theorists of the 19th century.

© The Author(s), under exclusive license to Springer-Verlag GmbH, DE, part of Springer Nature 2026

A. Fässler, *Mankind's Problem: Climate Change*,
https://doi.org/10.1007/978-3-662-71846-9_8

8.1 Wealth and poverty

From [43] by Christoph Rehmann-Sutter:[3] "The colonial conditions are being repeated in the climate crisis. Unlike the earlier colonizations, climate colonialism was not explicitly formulated as a political and cultural expansion program.

The climate effect is an unintended consequence. However, it was consciously accepted in existing industrial and consumer societies. We need to take a sober look at the colonial conditions that are repeated in the climate crisis. The cause of this is a development model that created prosperity in the industrialized countries. However, this is only possible at the expense of economically less developed countries.

Unilaterally distributed economic powers promote climate colonialism:

- Rich countries are offloading their pollution to societies with small footprints.
- The people of the industrialized countries also live at the expense of future generations.

Both aspects are morally untenable.

It is foreseeable and desirable that poor countries improve their standard of living (Human Development Index HDI). Air conditioning in hot southern countries is just as essential for a humane existence as is heating in cold regions.

The criteria of sustainability and climate justice can only be met by turning away from those system-immanent relations, namely that a higher standard of living causes a larger ecological footprint."

But the problem is not solved by net zero: The capacities for food and space are limited, as the Club of Rome established in its 1972 book *Limits to Growth!*

8.2 High standard of living versus population growth

On the one hand, there are the rich industrialized countries with their high standards of living and excessive consumption of resources, but a shrinking population (China's population is also declining). On the other hand, poor Third World countries have a modest consumption of resources per capita, but had a high reproduction rate in the last century (with Africa still having a fertility rate of 4.07 in 2025), which has led to a doubling of the population within the past 30 years.

---

[3] Philosopher and bioethicist at the University of Lübeck, Germany.

A controversial Swedish study from 2017 states that a child produces on average around 59 tons of CO_2e per year. For comparison: Not driving a car reduces emissions by around 2.4 tons per year. However, the study also takes into account the consequences of future generations.

As early as 1975, the Austrian-American social critic, philosopher, theologian and Catholic priest Ivan Illich dealt in detail with the two serious problem areas and proposed possible solutions in his book [27] entitled *Selbstbegrenzung, Eine politische Kritik der Technik* (English translation: *Blasphemy, A Radical Critique of Our Technological Culture*). He wrote: „**We all have to admit that there is an urgent need to limit procreation, consumption and waste production ...**"
Conclusion: Reducing the consumption of resources with the goal of net zero on the one hand and containing the world population on the other hand are necessary and sufficient conditions to make a livable human existence possible for future generations as well.

Small is beautiful: The motto „Bigger, Faster, More" has had its day. Nature groans under the exploitation of its resources. Overdeveloped technology and an unleashed economy are increasingly reaching their limits. At a time when society still unthinkingly adhered to the religion of „industrial gigantism", Ernst Schumacher (1911–1977)[4] and Sir Charles Chaplin (1889–1977) had already warned of the dangers of gigantism and already anticipated today's systemic crisis.
With Schumacher's vision of human-scale technologies that leave a smaller footprint[5] and allow people a maximum of self-determined activities, he anticipated much of what we understand today as sustainable development. His credo „Small is beautiful" is therefore more topical than ever, a perfect guide to a world in which the economy is ecologically oriented and serves people and not the other way around.

Gigantism also harbors enormous security risks and makes systems vulnerable on a grand scale. For instance, think of energy production (power plants, reservoirs, pipelines), huge data storage systems, oil tankers, container ships and airplanes. Decentralization and diversification, however, make a system robust.

---

[4] British economist of German origin.

[5] The complete opposite is provided by fantasies of technological omnipotence, such as tourist trips into space with hotel accommodation or to the Titanic in the depths of the ocean.

8.3 Contributions of various philosophers

If no lifespan is given, they are contemporaries. If no source is given, the article [40] was consulted.

1. **Alexander von Humboldt (1769–1859)** in [25]: „I could have concluded these considerations about the absorptivity and emissivity of the soil, on which the climate of the continents and the decrease in heat in the air generally depend, with an examination of the changes on the earth's surface which humans brought about by deforestation, by changing the distribution of the waters, and by producing great masses of steam and gas at the centers of industry. These changes are undoubtedly more essential than is generally assumed."

2. From the 1972 book **The Limits to Growth**: „Each year, exponential growth burdens the earth's ecosystem with several million additional people and billions of tons of waste. Even in oceans that once seemed inexhaustible, species after species of economically viable aquatic life are gradually being exterminated. Obviously, however, man learns nothing in his rush against the earthly borders."

3. **Hans Jonas (1903–1993)** developed in his 1979 book *Das Prinzip Verantwortung* (English translation: *The Imperative of Responsibility*), an ethic based on Immanuel Kant (1724–1804)[6], in which it becomes our duty to include future generations in our actions. His modern formulations of the categorical imperative are:
   - „Act so that the effects of your action are compatible with the permanence of genuine human life;
   - or expressed negatively: act so that the effects of your action do not disable such a life in the future;
   - or do not endanger the conditions for the indefinite continued existence of mankind."

   Jonas thus gave a decisive guideline for the mankind problem of climate change. The Golden Rule **Treat others the way you want to be treated** relates to the individual person. In contrast, the categorical imperative is about general legislation and consistency.

4. **Ivan Illich (1926–2002)** pleaded in his book [27] for **conviviality**. By this he means an autonomous and creative interpersonal relationship and the way people deal with their environment as opposed to the conditioned reactions of people to the demands of others and demands of an artificial world. To him, conviviality is individual autonomy that has an impact in the way people relate to each other.

---

[6] German philosopher of the Enlightenment and professor for logic and metaphysics in Königsberg, Prussia. His book *Critique of Pure Reason* marks the beginning of modern philosophy. He also wrote articles on astronomy and geosciences.

He believes that no amount of industrial productivity in society could really satisfy the needs it arouses among its members provided it falls below a certain level.

Technology should serve people and not the other way around. He continues: "Society is doomed when the growth of mass production causes the environment to become altogether inhospitable. Defective technology can make the world uninhabitable. ...In rich societies almost everyone is a destructive consumer."

Regarding mobility, he wrote decades ago: In recent years, advocates of progress had to admit that cars, the way they are used, are inefficient. Inefficient because one is obsessed with the idea that high speed means better transport.

As a starting point, he defines the **real speed** V as follows:

$$V = \frac{\text{displacement}}{\text{total time spent}}$$

The word **total** in the denominator includes more than just the pure driving time, as is shown in the following

Example of a small-car owner:

The costs refer to Swiss francs or US dollars or euros.

- Data:
  Duration of use 10 years, price 15,000 = 125/M (month), gross annual salary 90,000 = 7,500/M, driving distance 1,500 km/M. The car is de facto financed from the monthly net income of 3,500.
  With a monthly working time of 21 working days at 8 hours, the net income amounts to an hourly wage of $3,500/(8 \cdot 21) \approx 21$.
- Maintenance cost:
  - Insurance 500, vehicle tax 500, repairs 300, parking fee 500, tire change 500. In total 2,300/year.
  - Fuel cost 1.80/L (liters). Fuel consumption of 6.5 L/100 km results in an annual fuel cost of 2,106 for 18,000 km.
  - Acquisition cost 1,500/year
  In total 2,300+2,106+1,500 = 5,906/year = 492/M.
- Time spent per month:
  - At an average speed of 50 km/h for 1,500 km, this results in a travel time of 30 hours.
  - Monthly congestion time 7 h (i.e., 20 min/d).
  - Search for parking space 3.5 h (i.e., 10 min/d).
  - The working time for purchase and maintenance cost, where the purchase price is 125 /M and maintenance cost is 492 /M with the net hourly wage of 21, totals $(125 + 492)/21 = 29.5$ h.
  In total $(30 + 7 + 3.5 + 29.5)\,\text{h} = 70$ h/M.

Thus the real speed is

$$V = \frac{1,500\,\text{km}}{70\,\text{h}} = 21.5\,\text{km/h}.$$

It turns out that the plane performs well for very long distances, the train for somewhat longer distances and the bicycle for shorter distances.

Not to mention the resource consumption of the car (use of parking space and cultivated land, traffic jams, embodied energy) compared to trains and bicycles, as well as health and safety aspects.

We recommend reading the book [15]. It won the readers' prize of the 2022 German Business Book Prize.

5. **Hannah Arendt (1906–1975)** wrote about the student protests in her 1970 book *Macht und Gewalt* (English translation: *On Violence*). What was meant to be an analysis and criticism of the 1968 movement today reads like a description of Fridays for Future. Back then it was nuclear armament, today it is the climate crisis that makes the future a ticking time bomb for young people. It contains the following statements:

   „. . . The seemingly irresistible human progress, which in the course of the industrial revolution threatened only certain sections of the population with unemployment and triggered machine storming, today threatens the existence of entire popular groups and potentially that of humanity, in fact, of organic life in general. . . . What we are confronted with is a generation that is in no way certain that it has a future. . . . "

   Regarding the protests of young people that were already taking place at that time, she wrote in the book *Climate Crisis as a Generation Question*:

   „We are dealing with a new generation, a group of people who have the incredibly destructive tendencies of the rapid technical *progress* of the last decades in their flesh and blood. . . . This generation, compared to those *aged over thirty*, faces more consciously the Last Judgment that mankind is preparing for itself. This is only natural, not because they are younger, but because these were their first and defining experiences in the world."

6. **Ernst Ulrich von Weizsäcker and Anders Wijkman:** "We need a real fresh start. But this time we think it is necessary to deal with the philosophical roots of the dismal world situation. We must question the legitimacy of materialistic egoism, which is portrayed as the world's most powerful engine. The time is ripe for a new enlightenment, we think, or for other ways to replace today's short-term habits of thought and action."

7. **Christoph Lumer:**[7] "In the current climate policy, a fundamental moral principle is violated, namely the prohibition against harming others. If a sustainable climate policy is the best way to avoid climate damage, then

---

[7] Professor of moral philosophy at the University of Siena, Italy.

pursuing a sustainable climate policy is not just a question of possible political opportunities or the desire for a cleaner, more natural environment, but simply a moral duty."

8. **Kirsten Meyer**[8] writes in her book [35]: "Greenhouse gases, microplastics, deforestation – environmental damage of today is at the expense of humanity of tomorrow. This is immoral:
Future generations have a right to a good life. Dealing with natural resources is one of the most important ethical conflicts of the present. In this respect, we already have an obligation today to safeguard the interests of future generations, such as clean air, food, and satisfying the basic needs claimed by us. Two factors are decisive for the realization: the polluter-pays principle and the ability to finance it. Globally, the industrialized countries are largely the cause of the problem and are also in the best monetary position to do their part to fix it.
At the national level, the costs of environmental and climate protection measures should primarily be borne by the wealthy, who contribute most to the pollution of the environment by their way of life. An additional tax as an incentive to behave in a more environment-friendly way only makes sense if the state makes climate-friendly investments such as the expansion of local public transport and the railways."

9. **Bruno Latour (1947–2022)**[9] answered a series of questions in [40]. When asked whether the climate issue is also a generational issue, he replied: "The different generations do not have the same responsibility. That is why there is a reversal of the order of the generations, embodied by the young Swedish climate activist Greta Thunberg, which fascinates me: To the people of my generation who could have acted, she says: 'We young people are mature, in contrast to you. You are the childish ones, the immature ones'. Young people are not more mature than us, they come after us. Greta Thunberg is helping to process this new climatic situation. A truly prophetic figure. After all, a prophet does not care about the future, but about the present. . . . We have to come to terms with a situation that is unique in the history of earth."

10. **Wang Yangming (1472–1529):**[10]
"He who understands but fails to act, does not understand."

11. **John Broome:**[11]
"Most climate scientists are desperate. They've been doing their work for decades, and the world doesn't care."

---

[8] Professor at Humboldt University Berlin.

[9] was an important French sociologist and philosopher of science.

[10] Confucian philosopher in China's Ming Dynasty.

[11] Professor emeritus of Moral Philosophy at Oxford University. Worked on the 2014 IPCC Fifth Assessment Report.

Chapter 9
Appendix

9.1 Concept of a differential equation

First-order differential equations are equations for **functions** in which the first derivative occurs alongside the function itself.

Example 9.1 Exponential functions.
Let us look at functions $f(x)$ satisfying the differential equation

$$\frac{\mathrm{d}f}{\mathrm{d}x} = f.$$

If $f(x)$ assumes the value y, the differential equation states that the slope of the desired function f at point x must take on the value $y = f(x)$. Now we draw at a great number of points of the plane "compass needles" that have the required slope.

Along the straight line $y = 1$ the "compass needles" have the slope 1. Along the straight line $y = 1/2$ the "compass needles" have the slope $1/2$, and so on.

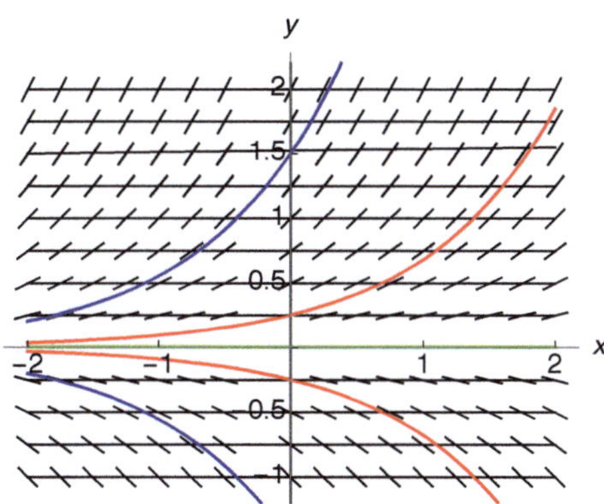

Obviously, for every **initial value** $f(0) = C$ there is exactly one solution, namely the **exponential function**

$$f(x) = C \cdot e^x, \quad \text{where } C \text{ is the initial value.}$$

The reason is that $f'(x) = (Ce^x)' = Ce^x = f(x)$.
For the special case $C = 0$, the solution is $f(x) = 0$.

Example 9.2 Solution to the Climate model

Consider the differential equation with given constants $a > 0$ and $b > 0$

$$\frac{\mathrm{d}T}{\mathrm{d}t} = a - b \cdot T$$

and the initial condition $T(0) = 0$.

The solution function T is given by $T(t) = \frac{a}{b}(1 - e^{-b \cdot t})$.

This is checked by inserting the solution function into the right-hand side of the differential equation:

$$a - b \cdot \frac{a}{b}(1 - e^{-b \cdot t}) = a \cdot e^{-b \cdot t},$$

which yields the left-hand side, namely the derivative $\frac{\mathrm{d}T}{\mathrm{d}t}$ of the solution function T. Furthermore, the initial condition is satisfied.

A particularly simple and physically interesting time-constant solution of the differential equation (often called the **stationary solution**) is obtained by equating the right-hand side to zero and solving for T. We then obtain $T = \frac{a}{b}$. The reason for this procedure is that the derivative of a constant function vanishes.

Such a procedure is interesting for more complicated non-linear differential equations. If a stationary solution exists, it can be computed by solving an ordinary equation.

9.2 Geometric sum and geometric series

By expanding the left-hand side one obtains the right-hand side:

$$(1 - m) \cdot (1 + m + m^2 + m^3 + \ldots + m^k) = 1 - m^{k+1}.$$

Hence

$$1 + m + m^2 + m^3 + \ldots + m^k = \frac{1 - m^{k+1}}{1 - m}.$$

If $|m| < 1$, then for $k \to \infty$ the term m^{k+1} converges to 0 and the geometric series (with infinitely many summands) becomes

$$1 + m + m^2 + m^3 \ldots = \frac{1}{1 - m}.$$

Example 9.3

$1 + \frac{1}{2} + \frac{1}{4} + \frac{1}{8} + \ldots = \frac{1}{1 - 1/2} = 2$, which is easily verified by a geometric consideration: We bisect the interval from 0 to 2, then bisect the remaining segment on the right half, etc. When bisecting continually, we can get as close to the number 2 as we like.

References

1. Binetoglou, I. und Susa, Z.: Die North Stream-Gaslecks im Vergleich (2022).
   https://www.catf.us/de/2022/10/putting-nord-stream-gas-leaks-perspective/
2. Blom, P.: Die Unterwerfung, Carl Hanser Verlag (2022).
3. Bressan, D.: Carbon Dioxide Peaked in 2022 At Levels Not Seen For Millions of Years,
   Science (2022).
4. Davies, Bethan: What is the volume of Thwaites glacier? in antarcticglaciers.org
   (2020).
   German Version: Bojanowski, A.: Entscheidet der Thwaites-Gletscher wirklich über
   das Schicksal der Menschheit?, Die Welt, Wissenschaft (2022).
   https://www.welt.de/wissenschaft/article236634497/Klimawandel-Der-Thwaites-
   Gletscher-als-angebliches-Untergangsorakel.html
5. Cai, W., Borlace, S., Lengainge, M. et al.: Increasing frequency of extreme El Niño
   events due to greenhouse warming. Nature Climate Change 4, 111–116 (2014).
   DOI: 10.1038/nclimate2100
6. Cai, W. et. al: Changing El Niño-Southern Oscillation in a warming climate. Nature
   Reviews Earth and Environment (2021).
   DOI: 10.1038/s43017-021-00199-z
7. Chandler D. L.: Shining brightly, MIT News (October 2011).
8. Cheng, L., Abraham, J., Trenberth, K.E. et al.: Another Year of Record Heat for the
   Oceans. Adv. Atmos. Sci. (2023).
   https://doi.org/10.1007/s00376-023-2385-2
9. Climate Change 2013: The Physical Science Basis. Fifth Assessment Report of the
   Intergovernmental Panel on Climate Change (IPCC).
10. IPCC, 2019: Summary for Policymakers. In IPCC Special Report on the Ocean
    and Cryosphere in a Changing Climate [H.-O. Pörtner, D.C. Roberts, V. Masson-
    Delmotte, P. Zhai, M. Tignor, E. Poloczanska, K. Mintenbeck, A. Alegria, M. Nicolai,
    A. Okem, J. Petzold, B. Rama, N.M. Weyer (eds.)].
    https://www.ipcc.ch/site/assets/uploads/sites/3/2021/12/SROCC-
    SPM_de_barrierefrei.pdf
11. IPCC Special Report on the Ocean and Cryosphere in a Changing Climate (2019).
    https://www.ipcc.ch/srocc/
12. IPCC Climate Change, The Physical Science Basis, Summary for Policymakers (2021)
    Contribution of Working Group I to the Sixth Assessment Report of the Intergov-
    ernmental Panel on Climate Change IPCC [Masson-Delmotte, V., P. Zhai, A. Pirani,
    S.L. Connors, C. Pean, S. Berger, N. Caud, Y. Chen, L. Goldfarb, M.I. Gomis, M.
    Huang, K. Leitzell, E. Lonnoy, J.B.R. Matthews, T.K. Maycock, T. Waterfield, O.
    Yelekçi, R. Yu, and B. Zhou (eds.)]. Cambridge University Press, Cambridge, United
    Kingdom and New York (2021).

https://www.ipcc.ch/report/ar6/wg1/downloads/report/IPCC_AR6_WGI_SPM_final.
pdf#page=33

13. Priyadarshi R. Sh. et al.: Climate Change 2022, Mitigation of Climate Change, Working Group III Contribution to the Sixth Assessment Report of the IPCC.
https://www.ipcc.ch/report/ar6/wg3/downloads/report/IPCC_AR6_WGIII_SPM.pdf

14. Grupp, M. et al.: IPCC Climate Change 2022: Mitigation of Climate Change, Working Group III.
https://www.ipcc.ch/report/sixth-assessment-report-working-group-3/

15. Diehl, K.: Autokorrektur–Mobilität für eine lebenswerte Welt, S. Fischer Verlag (2022).

16. Fässler, A.: Fast Track to Differential Equations, Applications-Oriented — Comprehensible — Compact. 2nd ed., Springer (2021).

17. Fässler, A.: Unsolved: Climate Change Problem, Journal of Earth and Environmental Science Research, Vol.7, ISSN: 2634-8845 (2025).
https://www.onlinescientificresearch.com/journals/jeesr/archives/2025/7/8

18. Fässler, A.: Groups, Symmetry and Symmetry Breaking in the Proceedings on Mathematical Modelling in Education and Culture: ICTMA 10, p. 143–152 (2003) of the ICTMA 10 Conference at Tsinghua University in Beijing (2002), Horwood Publishing Limited, 2003, ISBN 1-904275-05-2.

19. Farinotti, D. et al.: A consensus estimate for the ice thickness distribution of all glaciers on Earth, Nature Geoscience 12, 168–173 (2019).
https://www.slf.ch/de/newsseiten/2019/02/eisvolumen-aller-gletscher-der-welt-neu-berechnet.html

20. Forster, P. M. et al.: Indicators of Global Climate Change 2024: annual update of key indicators of the state of the climate system and human influence, Earth System Science Data,(June 2025).
https://doi.org/10.5194/essd-17-2641-2025

21. Friedlingstein, P. et al.: Global Carbon Budget 2024, Earth System Science Data, Volume 17, issue 3, (2025) 965–1039. https://doi.org./10.5194/essd-17-965-2025.

22. Stewart, I. and Golubitsky, M.: Fearful Symmetry: Is God a Geometer?, Dover Publications Inc., New York (2010), Copyright 1992.

23. Goosse, H.: Climate System Dynamics and Modelling, Cambridge Univesity Press (2015).

24. Gravens, H. D.: Impact of Fossil Fuel Emissions on Atmospheric Radiocarbon over this Century (2015).

25. Humboldt, A.: Central-Asien, Untersuchungen über die Gebirgsketten und die vergleichende Klimatologie, p.214 (1844).

26. Huss, M. and Farinotti, D.: Distributed ice thickness and volume of all glaciers around the globe. Journal of Geophysical Research, 117, F04010 (2012).
doi:10.1029/2012JF002523

27. Illich, I.: Blasphemy: A Radical Critique of Our Technological Culture. Morristown, NJ, Aaron Press, July 1995.

28. Jonas, H.: Das Prinzip Verantwortung (1979).

29. Jones, Chris D.; Friedlingstein, Pierre: Quantifying process-level uncertainty contributions to TCRE and carbon budgets for meeting Paris Agreement climate targets. Environmental Research Letters, Volume 15, Number 7 (2020): 074019.
doi.org/10.1088/1748-9326/ab858a

30. Klima, I.: Liebe und Müll, Hanser Verlag (1991).

31. Lanz, K.; Müller, L.; Rentsch, C.; Schwarzenbach, R.: For Climate's Sake, in collaboration with the Department of Environmental Sciences, Swiss Federal Institute of Technology Zurich ETHZ. Lars Müller Publishers (2011). A recommended book on climate change with numerous impressive photos.

32. Levin, I. et al.: Observations and modelling of the global distribution and long-term trend of atmospheric $^{14}CO_2$,Tellus, Ser. B, Chem. Phys. Meteorol., 62(1), 26–46, (2010).
doi:10.1111/j.1600-0889.2009.00446.x.(2010)

33. Lindsey, R.: Climate Change: Atmospheric Carbon Dioxide (2022).
    http://www.climate.gov/news-features/understanding-climate/climate-change-atmospheric-carbon-dioxide
34. Mackintosh, A.: Thwaites Glacier and the bed beneath, Nat. Geosci. 15, 687-688 (2022).
    https://doi.org/10.1038/s41561-022-01020-2
35. Meyer, K.: Was schulden wir künftigen Generationen? Herausforderung Zukunftsethik, Reclam (2018).
36. Meadows, D. et al: Limits to Growth (1972).
37. Hogonnet, R. et al.: Accelerated global glacier mass loss in the early twenty-first century, Nature 592, p.726–731 (2021).
    https://www.nature.com/articles/s41586-021-03436-z
38. Niedermair, M., et al.: Klimawandel & Artenvielfalt. Wie klimafit sind Österreichs Wälder, Flüsse und Alpenlandschaften?, (2007).
39. Ortner, G., Bründl, M., Kropf, C. M., Bühler, Y., and Bresch, D. N.: Climate change impacts on snow avalanche risk in alpine regions, EGU General Assembly 2022, Vienna, Austria, May 2022, EGU22-8571.
    https://doi.org/10.5194/egusphere-egu22-8571, 2022
40. Interviews und Beiträge zur Klimakrise, Philosophie Magazin, Sonderausgabe 16, Herbst/Winter 2020 / 2021, Philomagazin Verlag GmbH.
41. Oroschakoff, K.: Planet A of NZZ of 25 January 2023.
    https://img.nzz.ch/S=W972/O=75/https://img.nzz.ch/2023/01/25/d84bcd9a-5fdd-4639-8398-59e79158ed09.png
42. Paleoclimate Working Group of the National Center for Atmospheric Research NCAR in Boulder, Colorado, supported by the National Science Foundation NSF, (2017).
43. Rehmann-Sutter, C.: Klimawandel – und die Philosopie? (2019).
    https://www.philosophie.ch/2019-07-16-rehmannsutter
44. Rignot, E. et al.: Four decades of Antarctic Ice Sheet mass balance from 1979–2017, PNAS National Academy of Sciences (2019).
45. Ringenbach, A.: Gletscher ein Markenzeichen der Schweiz – wie lange noch?, Maturaarbeit Kantonsschule Solothurn, Betreuer: B. Marti (2014).
46. B. D. Santer et. al.: Exceptional stratospheric contribution to human fingerprints on atmospheric temperature. Proceedings of the National Academy of Sciences (2023).
    https://www.pnas.org/doi/10.1073/pnas.2300758120
47. Rounce A. et al., Global glacier change in the 21st century: Every increase in temperature matters, Science 379, 78–83 (2023).
48. Schmidt, B.E., Washam, P., Davis, P.E.D. et al.: Heterogeneous melting near the Thwaites Glacier grounding line. Nature 614, 471–478 (2023).
    https://doi.org/10.1038/s41586-022-05691-0
49. The International Thwaites Glacier Collaboration: Thwaites Glacier, Seafloor images explain Thwaites Glacier retreat (September 2022).
    https://thwaitesglacier.org/news/seafloor-images-explain-thwaites-glacier-retreat
50. Shi, J-R. et.al.: Ocean warming and accelerating Southern Ocean zonal flow, Nature Climate Change, vol.11, 1090–1097 (2021).
    https://doi.org/10.1038/s41558-021-01212-5
51. Statista Research Department: Bevölkerungsdichte nach Kontinenten 2021 und 2100. de.statista.com (2022).
52. Muschter, R.: Bevölkerungsdichte in Indien bis 2050, Statista: de.statista.com (2023).
53. Muschter, R: Gesamtbevölkerung in Indien, Statista: de.statista.com (2023).
    https://de.statista.com/statistik/daten/studie/19326/umfrage/gesamtbevoelkerung-in-indien/
54. Muschter, R: Bevölkerungsdichte in China bis 2050, Statista: de.statista.com (2023).
55. Statista Research Department: CO_2-Ausstoss weltweit bis 2021 (November 2022).
    https://de.statista.com/statistik/daten/studie/37187/umfrage/der-weltweite-co2-ausstoss-seit-1751/

56. Tierney, J. E. et al.: Past Climate Change inform our Future, Science, vol. 370, issue 6517, eaay 3701, (2020).
    https://doi.org/10.1126/science.aay3701
57. Thwaites-Gletscher: Kipp-Punkt fürs Klima, Video NZZ (2022).
    https://cdn.jwplayer.com/previews/E03fBDgL
58. Zhang, Z., Poulter, B., Feldmann A.F. et al.: Recent intensification of wetland methane feedback. Nature Climate Change (2023).
    https://doi.org/10.1038/s41558-023-01629-0

Internet only:

59. https://berkeleyearth.org/global-temperature-report-for-2024/
60. https://www.meereisportal.de/newsliste/detail/antarktische-meereisausdehnung-erreicht-allzeittief. (2023)
61. https://nsidc.org/arcticseaicenews/charctic-interactive-sea-ice-graph/
62. https://wiki.bildungsserver.de/klimawandel/index.php/Arktisches_Meereis
63. Carbon The Unauthorised Biography: https://www.thecarbonmovie.com
    Video in german: https://www.spektrum.de/video/kohlenstoff-eine-geschichte-von-leben-und-tod/2007691
64. Levin I., et al. of Carbon Cycle Group, University of Heidelberg: Long-term monitoring of $^{14}CO_2$ in the global atmosphere, (2022).
    https://www.iup.uni-heidelberg.de/de/research/kk
65. graph of mean tmperature curve on the earth's surface to be found under
    https://de.wikipedia.org/wiki/Globale Erwärmung
66. Internet under Earth.org.
67. https://de.wikipedia.org/wiki/Liste_der_grössten_Methanemittenten
68. https://www.nasa.gov/feature/goddard/2017/messier-74
69. Rabo, O.: Cooler Future under
    https://www.coolerfuture.com/blog-de/co2-aquivalent
70. Temperaturanomalien von 1850 bis 2020 under
    https://www.metoffice.gov.uk/hadobs/hadcrut5/index.html
71. Aktuelle Klimaänderungen under
    https://wiki.bildungsserver.de/klimawandel/index.php/Aktuelle_Klimaänderungen
72. Skizze Thwaites-Gletscher under
    https://www.sciencealert.com/scientists-discover-what-s-underneath-the-unstable-thwaites-glacier
73. Arktis ohne Eis?, WWF (2020).
    https://www.wwf.de/themen-projekte/projektregionen/arktis/arktis-ohne-eis
74. State of the Global Climate, World Meterological Organization WMO, March 2025.
75. Climate Change 2022, Mitigation of Climate Change, Working Group III Contribution to the Sixth Assessment Report of the Intergovernmental Panel on Climate Change, Summary for Policymakers (2022).
    https://www.ipcc.ch/report/ar6/wg3/downloads/report/IPCC_AR6_WGIII_SPM.pdf

Index